6

AF771956

WASTE AND WASTE MANAGEMENT

Additional books in this series can be found on Nova's website under the Series tab.

Additional e-books in this series can be found on Nova's website under the e-book tab.

WASTE AND WASTE MANAGEMENT

POTENTIALLY CONTAMINATED SITES MANAGED BY USDA AND DEPARTMENT OF THE INTERIOR

IDENTIFICATION, FUNDING AND LIABILITY ISSUES

WASTE AND WASTE MANAGEMENT

POTENTIALLY CONTAMINATED SITES MANAGED BY USDA AND DEPARTMENT OF THE INTERIOR

IDENTIFICATION, FUNDING AND LIABILITY ISSUES

ROGER L. BROOKS
EDITOR

nova publishers
New York

Library of Congress Cataloging-in-Publication Data

ISBN: 978-1-63483-076-8

Published by Nova Science Publishers, Inc. † New York

CONTENTS

PREFACE

USDA and Interior manage over 600 million acres of land, including sites contaminated from prior uses or events, such as mining or toxic spills. These lands are managed by five Interior agencies, including BLM and the National Park Service, and five USDA agencies, including the U.S. Forest Service. These agencies must identify and report to EPA certain facilities that may threaten human health or the environment and, under some circumstances, clean them up. They must also report cost estimates for addressing contamination at certain sites, called environmental liabilities. This book examines USDA's and Interior's efforts to identify these sites, funding to address the sites, reported environmental liabilities, and EPA's role in addressing the sites. Furthermore, this book provides a summary of the Superfund Amendments and Reauthorization Act and related provisions of the Act.

In: Potentially Contaminated Sites …
Editor: Roger L. Brooks
ISBN: 978-1-63483-076-8

Chapter 1

HAZARDOUS WASTE: AGENCIES SHOULD TAKE STEPS TO IMPROVE INFORMATION ON USDA'S AND INTERIOR'S POTENTIALLY CONTAMINATED SITES*

United States Government Accountability Office

WHY GAO DID THIS STUDY

USDA and Interior manage over 600 million acres of land, including sites contaminated from prior uses or events, such as mining or toxic spills. These lands are managed by five Interior agencies, including BLM and the National Park Service, and five USDA agencies, including the U.S. Forest Service. These agencies must identify and report to EPA certain facilities that may threaten human health or the environment and, under some circumstances, clean them up. They must also report cost estimates for addressing contamination at certain sites, called environmental liabilities.

GAO was asked to review the status of USDA's and Interior's potentially contaminated sites. This report examines USDA's and Interior's (1) efforts to identify these sites, (2) funding to address the sites, (3) reported environmental liabilities, and (4) EPA's role in addressing the sites. GAO reviewed relevant

* This is an edited, reformatted and augmented version of a United States Government Accountability Office publication, No. GAO-15-35, dated January 2015.

laws and government accounting standards; examined agencies' policies, site inventory data from September 2013 to July 2014, and financial statements; and interviewed EPA, Interior, and USDA officials.

WHAT GAO RECOMMENDS

GAO recommends that USDA develop plans and procedures for completing its site inventories and that EPA clarify which USDA and Interior sites need an environmental assessment. Interior and EPA generally agreed with GAO's findings. USDA disagreed that its incomplete inventory affects the effectiveness of its cleanup programs, but GAO continues to believe that effective program management requires reliable data.

WHAT GAO FOUND

The U.S. Department of Agriculture (USDA) and the Department of the Interior (Interior) have identified many contaminated and potentially contaminated sites, but neither agency has a complete inventory. As of April 2014, USDA had identified 1,491 contaminated sites and many potentially contaminated sites, including landfills and shooting ranges. However, USDA does not have a reliable, centralized site inventory or plans and procedures for completing one, in particular, for abandoned mines. Without a reliable inventory or plans and procedures for developing one, USDA cannot effectively manage its cleanup programs. As of April 2014, Interior had an inventory of 4,722 sites with confirmed or likely contamination. In addition, Interior's Bureau of Land Management (BLM) had identified over 30,000 abandoned mines that were not yet assessed for contamination, and this inventory is not complete. BLM is working to improve the completeness and accuracy of its inventory.

In fiscal year 2013, USDA allocated over $22 million for environmental cleanup efforts, and Interior allocated over $13 million. Specifically, in fiscal year 2013, USDA allocated about (1) $3.7 million for cleanup projects department-wide, primarily for one large site; (2) $4.3 million to mitigate contamination at grain silos and foreclosure properties; and (3) approximately $14 million for the U.S. Forest Service to assess and clean up sites. In fiscal year 2013, Interior allocated almost $10 million for cleanup projects department-wide, the National Park Service allocated an additional $2.7 million and the Fish and Wildlife Service allocated over $800,000 for environmental assessment and cleanup projects. In addition, BLM allocated more than $34 million to its hazardous waste management and abandoned mine programs.

Both USDA and Interior report probable and reasonably estimable environmental liabilities on their financial statements. In fiscal year 2013, USDA reported environmental liabilities of $176 million, most of which was for asbestos cleanup. Interior reported environmental liabilities of $192 million to address 434 sites. Interior also disclosed that it will incur $62 million to $139 million in cleanup costs at government-acknowledged sites—sites with damage caused by nonfederal entities. The majority of the costs Interior disclosed were for cleanup activities at 85 abandoned mines. As USDA and Interior complete their inventories and assess them for contamination, it might be expected that related environmental costs for both agencies will increase as additional sites are considered.

The Environmental Protection Agency's (EPA) role with regard to USDA and Interior sites is to maintain a list of potentially contaminated sites based on data from these agencies, take steps to assure that the agencies assess these sites for contamination, determine whether the sites should be proposed for the National Priorities List, and oversee certain cleanup activities. EPA has compiled a docket list of 260 USDA sites and 528 Interior sites that may pose health or environmental risks, most of which have been assessed. Disagreement with USDA and Interior over their need to assess the remaining sites and differing information on whether this requirement has been met at some sites means EPA cannot assure that the assessments are conducted in a timely manner and, ultimately, that sites most in need of remediation are addressed.

ABBREVIATIONS

AML	Abandoned Mine Lands
APHIS	Animal and Plant Health Inspection Service
ARS	Agricultural Research Service
BLM	Bureau of Land Management
CCC	Commodity Credit Corporation
CEE	Center for Environmental Excellence
CERCLA	Comprehensive Environmental Response,

	Compensation, and Liability Act
CHF	Central Hazardous Materials Fund
Corps	U.S. Army Corps of Engineers
DOD	Department of Defense
ECAP	Environmental Compliance and Protection
EDL	Environmental and Disposal Liability
EMD	Environmental Management Division
EPA	Environmental Protection Agency
FASAB	Federal Accounting Standards Advisory Board
FSA	Farm Service Agency
FUDS	Formerly Used Defense Sites
HMMA	Hazardous Materials Management Appropriation
HRS	Hazard Ranking System
Interior	Department of the Interior
NDAI	no DOD action indicated
NFS	National Forest System
NPL	National Priorities List
NPS	National Park Service
OIG	Office of Inspector General
OMB	Office of Management and Budget
PRP	potentially responsible party
RCRA	Resource Conservation and Recovery Act
SFFAS	Statement of Federal Financial Accounting Standards
USDA	U.S. Department of Agriculture

* * *

January 16, 2015

The Honorable Paul D. Tonko
Ranking Member
Subcommittee on Environment and the Economy
Committee on Energy and Commerce
House of Representatives

Dear Mr. Tonko:

Agencies within the U.S. Department of Agriculture (USDA) and the Department of the Interior (Interior) manage over 700 million acres of land that include national parks, national forests, research centers, and laboratories. This vast acreage includes land that may be contaminated with hazardous waste[1] from prior uses or events, such as landfills or toxic spills. In addition, abandoned mines from private mining activities on Interior and USDA lands can present major environmental cleanup challenges and expenses for the federal government. These lands are managed by five agencies within Interior, including the Bureau of Land Management (BLM) and the National Park Service (NPS), and five agencies within USDA, including the Forest Service and the Agricultural Research Service.

The Comprehensive Environmental Response, Compensation, and Liability Act (CERCLA) of 1980 requires the Environmental Protection Agency (EPA) to compile a list of contaminated and potentially contaminated federal facilities. This list, known as the Federal Agency Hazardous Waste Compliance Docket (docket) is based on information that agencies are required by CERCLA and the Resource Conservation and Recovery Act (RCRA) of 1976 to report to EPA. In addition, the Chief Financial Officers Act of 1990[2] requires and Office of Management and Budget (OMB) Circular A-136 and the Federal Accounting Standards Advisory Board's (FASAB) Statement of Federal Financial Accounting Standards (SFFAS) No. 53 direct federal agencies to include environmental liabilities in their annual financial statements.

In 1987 and 1994, GAO found that Interior and USDA were slow to complete an inventory of potentially contaminated sites for which they were responsible because of the size of their landholdings and limited resources. [4] GAO's past work has also shown that BLM and Forest Service, the two principal agencies that manage federal lands available for hardrock mining, had no definitive estimates of the number of abandoned mines on their lands.[5]

You asked us to review the status of USDA's and Interior's potentially contaminated sites. This report examines (1) USDA's and Interior's efforts to develop an inventory of their contaminated and potentially contaminated sites, (2) funding levels for addressing their contaminated sites, (3) the amount of environmental liabilities the agencies report on their financial statements, and (4) EPA's role with regard to these sites.

To conduct this work, we reviewed relevant inventory data covering the period of September 2013 to June 2014 from 10 agency sources and budget data related to environmental cleanup projects for fiscal years 2003 through 2013. We also reviewed department and agency documents; government accounting standards and laws; and interviewed EPA, USDA, and Interior officials. To determine the status of USDA's and Interior's efforts to identify the universe of potentially contaminated sites on lands they manage, we reviewed each agency's policies and guidance on identifying new sites, and we interviewed agency officials on the completeness of their inventories. We collected data on both the number of contaminated sites in each agency's inventory, as well as the number of sites each agency has identified that could potentially be contaminated but have not yet been assessed to confirm contamination. Agencies with small inventories provided site numbers or a list. For the other agencies with more extensive inventories, we obtained information from a number of databases. To assess the reliability of the inventory lists from the agency databases, we reviewed available documents to determine the source of the information, data entry steps, and the completeness of the inventory, and interviewed agency officials. We found that some of the databases had inaccuracies and were not complete, but we determined that the data systems represented the available inventory identified and used by the agency and, therefore, were sufficiently reliable for our purposes of reporting USDA's and Interior's known contaminated sites. However, we do not feel that USDA's data on potentially contaminated sites was sufficiently reliable for the purpose of reporting the number of known potentially contaminated sites. To determine the funding levels for addressing the agencies sites in recent years, we selected the last decade as a reasonable period for analysis and, therefore, we collected and reviewed budget data for fiscal years 2003 through 2013. We also interviewed agency officials regarding estimated costs to address the abandoned mines on these agencies' properties. To determine the amount of environmental liabilities Interior and USDA report, we reviewed these departments' financial statements and interviewed officials in USDA's and Interior's Offices of Chief Financial Officer, USDA's Office of General Counsel, and Interior's Office of Solicitor. In addition, we collected and reviewed department and agency guidance on environmental liabilities and cost estimating. At Interior, we reviewed information from the Environmental and Disposal Liability (EDL) database. We reviewed the Chief Financial Officers Act of 1990, OMB Circular A-136, and the Federal Accounting Standards Advisory Board's (FASAB) Statement of Federal Financial Accounting Standards (SFFAS) No. 5 to better understand the applicable criteria.[6] To determine the role of EPA with regard to these sites, we interviewed EPA officials with the Federal Facilities Restoration and Reuse Office and reviewed data from EPA's Comprehensive Environmental Response, Compensation, and Liability Information System on the status of USDA and Interior sites, as of December 2013. To assess the reliability of EPA's data on the status of the sites, we also obtained information from

USDA and Interior on the status of their sites and documented any differences; we found the data to be sufficiently reliable for the purposes of this report. Finally, for all four objectives, we reviewed relevant laws—in particular, CERCLA and RCRA—and associated regulations and directives. Appendix I provides a detailed description of our objectives, scope, and methodology.

We conducted this performance audit from April 2013 to January 2015 in accordance with generally accepted government auditing standards. Those standards require that we plan and perform the audit to obtain sufficient, appropriate evidence to provide a reasonable basis for our findings and conclusions based on our audit objectives. We believe that the evidence obtained provides a reasonable basis for our findings and conclusions based on our audit objectives.

Background

Federal law requires agencies to identify and report certain facilities where there is or has been hazardous waste activity or a release of hazardous substances. In addition, federal accounting standards direct federal agencies to develop cost estimates for certain environmental liabilities and report those estimates on their financial statements.

Federal Agency Inventory Requirements

Federal agencies began developing inventories of contaminated and potentially contaminated sites in response to laws passed in the 1980s. In 1980, Congress enacted CERCLA to respond to problems caused by improper disposal of hazardous substances in the past. Under CERCLA, responsible parties are liable for conducting or paying for site cleanup of hazardous substances, or for reimbursing others who conduct cleanups on their behalf, including federal agencies.[7] Furthermore, by June 1981, all such facility owners and operators were required to notify EPA of the existence of their facilities, as well as known, suspected, or likely releases of hazardous substances. These liability and notification provisions applied to the federal government and its contractors and lessees as well.

Beginning in 1986, RCRA required agencies to undertake a continuing program to identify and report to EPA those facilities that the agencies own or operate—or previously owned or operated—that involved the treatment, storage, or disposal of hazardous wastes. These inventories must be updated biennially. RCRA does not set a deadline for completing the inventories.

Reporting Environmental Liabilities on Agency Financial Statements

Federal accounting standards direct agencies responsible for cleaning up contamination to estimate cleanup costs and report those estimates on their financial statements as environmental liabilities. The Federal Accounting Standards Advisory Board's (FASAB) Statement of Federal Financial Accounting Standards (SFFAS) No. 58 outlines standards based on whether events are government-related or government-acknowledged. Government-related events, or sites, involve interaction between federal entities and their environment, and include hazardous waste spills on federal property caused by federal operations. Government-acknowledged events are sites that are of financial consequence to the federal government because it chooses to respond to the event, and include toxic waste damage caused by nonfederal entities. SFFAS No. 5 directs an agency to report as a liability on its financial statements an amount for probable and reasonably estimable future costs associated with cleaning up contamination associated with these events, and to disclose the nature of the contingency, an estimate of the possible liability or the range of the possible liability, or a statement that such an estimate cannot be made for reasonably possible future costs in the notes to the financial statements.

Guidance for implementing the standards[9] indicates that there are two ways a federal agency's environmental liabilities may meet the "probable" cost criteria. The first way is when contamination is known, government-related, and represents a legal liability. The second way is when the contamination is known, government-acknowledged, and funds have been appropriated for a cleanup or an event has occurred causing another party to expect payment (e.g., a contractor has performed cleanup of a site).

The "reasonably estimable" criteria relates to an agency's ability to reliably estimate possible future costs associated with the cleanup of a contaminated site. In making such estimates, the agency is to consider several key factors, such as whether remedial investigations or feasibility studies have been conducted, prior experience with similar sites or conditions, and the availability of technology required to conduct cleanup operations.

If a liability is probable, but a reasonable estimate of costs can only be expressed as a range, SFFAS No. 5 directs the agency to report the low end of the range on the face of the financial statement and disclose the complete range in the notes of its financial statement.[10] Similarly, if a liability is "reasonably possible"—that is, the probability of future expenditures for a given site is less than probable but greater than remote—then SFFAS No. 5 directs disclosure of the range of possible liability or a statement that such an estimate cannot be made.

USDA and Interior Have Identified Thousands of Contaminated and Potentially Contaminated Sites, but Neither Agency Has a Complete Inventory of Abandoned Mines

USDA and Interior have identified many contaminated and potentially contaminated sites[11], but neither agency has a complete inventory of sites, in particular, abandoned mines. There are approximately 1,491 USDA sites with confirmed contamination. However, USDA has not compiled a reliable, centralized inventory of known sites that are potentially contaminated and does not have plans or procedures for developing one. In particular, considerable work remains to identify abandoned mines on National Forest System (NFS) lands.[12] Interior has identified 4,722 sites with confirmed or likely contamination on land it manages. Interior officials believe that this inventory is largely complete, except for abandoned mines. In this regard, Interior's BLM has over 30,000 abandoned mines in its inventory that it has not yet assessed for contamination, but the inventory is not complete. For example, BLM has estimated that there may be about 100,000 additional sites in California, Nevada, and Utah that have not yet been located, included in the inventory, and assessed for contamination. BLM has a centralized inventory database of its sites and has been taking steps to improve its data quality and completeness.

USDA

According to our analysis of several USDA data sources,[13] as of April 2014, there were at least 1,491 contaminated sites on land currently or previously managed by USDA. These sites include 1,422 Forest Service sites—primarily abandoned mines—2 Animal and Plant Health Inspection Service (APHIS) sites, 3 Agricultural Research Service (ARS) sites, 61 former grain storage sites once managed by Commodity Credit Corporation (CCC), and 3 foreclosure properties belonging to the Farm Service Agency (FSA).[14]

In addition to sites with confirmed contamination, USDA agencies have also identified some potentially contaminated sites. ARS has identified 3 sites that are potentially contaminated. The Forest Service maintains multiple inventories of potentially contaminated sites that include landfills, shooting ranges, and cattle dip vats. However, there is no centralized database of these sites and no plans or procedures for developing one; the inventories are maintained at the regional level. A number of Forest Service officials we spoke with believe that most potentially contaminated sites in these categories have been identified but, for the reasons discussed below, they were unable to provide us with inventory numbers for all these sites.

These various inventories do not provide a complete picture of the extent of USDA's potentially contaminated sites. For example, there are an unknown number of potentially contaminated former grain storage sites in the 29 states where the CCC previously used carbon tetrachloride. This number is unknown because the CCC relies on the states to notify it of potential contamination, and 25 of the 29 states have not yet reported whether there is suspected contamination at their former CCC grain storage sites. The Forest Service also deals with various other types of hazardous waste sites, such as methamphetamine laboratories, roadside spills, and waste dumps. Forest Service officials said that, since these types of sites may involve illegal activities and are, therefore, not routinely reported, it is not possible to develop a comprehensive inventory of these types of sites.[15]

In addition, the Forest Service has not developed a complete, consistent, or usable inventory of abandoned mines and has no plans and procedures for developing such an inventory. The Forest Service estimates that there are between 27,000 and 39,000 abandoned mines on its lands[16]—approximately 20 percent of which may pose some level of risk to human health or the environment, based on the professional knowledge and experience of its staff. Such risks may include chemicals and explosives, acid mine drainage, and heavy metal contamination in mine waste rock. However, because the Forest Service does not have a complete inventory of abandoned mine sites, the actual number of abandoned mines on NFS lands is unknown. Forest Service officials told us that they do not have the resources to complete a comprehensive inventory of all potentially contaminated abandoned mines on the agency's lands. However, without a comprehensive inventory of such sites or plans and procedures for developing one, USDA and the Forest Service will not have reasonable assurance that they are prioritizing and addressing the sites that pose the greatest risk to human health or the environment.

The universe of abandoned mines on NFS lands is unknown for several reasons. According to a USDA official, USDA attempted to create a national inventory of NFS mines in 2003, with the creation of the Center for Environmental Excellence (CEE) database. This inventory pulled data from regional Forest Service inventories and currently shows 20,401 potentially contaminated abandoned mine or mill sites. 17 However, these data are not complete or current and are therefore not useable for purposes of identifying a complete and accurate number of potentially contaminated abandoned mines currently on NFS lands. The Forest Service established the Abandoned Mine Lands (AML) database in 2008 to aggregate all available data on abandoned mines on NFS lands. The AML database drew data on pending abandoned mine sites from the CEE database and Forest Service regional inventories, as well as from the U.S. Geological Survey and various other federal, state, and local databases. USDA officials said that, once the AML database was established, the purpose of CEE shifted away from maintaining an AML inventory to tracking sites that entered into the CERCLA process.[18]

However, the AML database also has a number of shortcomings. For example, the data migration from multiple inventories led to data redundancy issues, such as some mine sites being listed multiple times under the same or different names. In addition, USDA officials told us that there is a lot of variation in the accuracy and completeness of the data on these mine sites, but a quality assurance review has not yet been performed. One Forest Service official said that, because of these problems, the data in the AML database are currently unusable for purposes of compiling a complete and accurate inventory of abandoned mines. In 2012, the Forest Service tried to obtain agency resources necessary to clean up the database. Even though the Forest Service rated this project as "critical," the project did not receive any resources because other projects were deemed more important, according to a Forest Service official.

Similarly, there are several problems with the Forest Service's regional abandoned mine inventories.[19] First, some regional inventories are incomplete. For example, officials in Forest Service Region 10, which is composed solely of the state of Alaska, believe that their inventory of active mine sites is largely complete, but they said that there may be some abandoned mines scattered throughout Tongass and Chugach National Forests that have not yet been inventoried. They said that Forest Service Region 10 does not have enough staff to assess all abandoned mines across such a large area.[20]

Second, several Forest Service regional inventories contain inaccurate data. For instance, Forest Service Region 1's inventory data include all mine records from the state of Montana—including those not on Forest Service lands. Forest Service Region 8 maintains several different abandoned mine databases, and one official told us that there are multiple redundancies across these databases.

Third, the Forest Service's regional offices maintain their inventories differently. Some regional offices maintain their own inventories of potentially contaminated sites, whereas other regional offices utilize state or local agencies' inventories. In at least one instance, these inventories were not easily accessible. Specifically, Forest Service Region 2 relies heavily on various state and local inventories, but officials in that region did not provide us with data from those inventories when we requested them, instead they suggested we contact the state agencies.[21]

Finally, the type of data on abandoned mines varies from region to region and is therefore difficult to consolidate into a coherent national database. Some regional offices track mines at the site level, some by their features—such as mine shafts, pits, ore piles, or machinery—and some use both approaches.[22] For example, officials in Forest Service Region 3 told us that they have identified over 3,000 abandoned mine sites, whereas officials in Forest Service Region 4 told us that they have identified approximately 2,000 mine features but have not yet consolidated these features into mine sites.

Despite the lack of a complete inventory of potentially contaminated sites, the Forest Service said that the highest risk sites are well-known due to their historic nature and past production of metal resources. According to the Forest Service, the abandoned mines that have not yet been identified include many small, poorly documented mines. The Forest Service said that many of these mines did not ever produce any ore and thus were only exploratory mines. Because of the high risk potential, the abandoned mine sites with the greatest human visitation and/or sensitive ecological receptors have been given priority for assessment, according to Forest Service officials. These officials said that only sites with significant risk are funded.

Interior

Interior has identified 4,722 sites with confirmed or likely contamination. These include 4,098 BLM sites that the agency reports have confirmed contamination or require further investigation to determine whether remediation is warranted. The majority of these sites are abandoned mines. NPS has identified 417 sites with likely or confirmed contamination, the Bureau of Indian Affairs, 160; the Fish and Wildlife Service, 32; and the Bureau of Reclamation, 15. These agencies have identified additional locations of concern that require verification or initial assessment to determine if there are environmental hazards at the sites. Except for BLM, agency officials told us that they believe they have identified all sites with likely environmental contamination. The total number of sites BLM may potentially have to address is unknown, due primarily to incomplete and inaccurate data on abandoned mines on land managed by the agency.

BLM accounts for the largest number of contaminated sites and sites that need further investigation in Interior's inventory. Table 1 shows the number of contaminated or potentially contaminated sites in BLM's inventory as of April 2014.

Table 1. BLM's Contaminated or Potentially Contaminated Sites Identified as of April 2014

	Hazardous materials sites	Abandoned mines with potential environmental contamination	Total sites
Requires further investigation and/or remediation	346	2,853	3,199
Has remediation measures planned or under way	546	353	899
Has been remediated	1373	886	2,259
Requires no further action	88	568	656
Total	**2,353**	**4,660**	**7,013**

Source: Bureau of Land Management's (BLM) Abandoned Mine Lands/Site Cleanup Inventory database. | GAO-15-35

Note: According to BLM, the Abandoned Mine Lands/Site Cleanup Inventory database is the agency's source of information regarding the inventory and status of abandoned mine and hazardous materials sites. BLM has not yet assessed all the sites on the ground, and the agency is constantly reviewing and updating the data.

As table 1 shows, BLM identified 2,353 hazardous materials sites as of April 2014, including underground storage tanks or landfills. Of these 2,353 sites, 892 sites may still need to be addressed; 546 sites have confirmed contamination with remediation measures planned or under way, and 346 sites may potentially be contaminated and require an initial or further assessment to determine whether remediation is warranted.[23]

In addition, BLM identified 45,312 abandoned mine sites in its inventory database. As of April 2014, BLM identified that 4,660 abandoned mine sites have or likely have environmental contamination. Of these 4,660 environmentally contaminated mine sites, 353 have remediation measures planned or under way, and 2,853 require further investigation to determine whether remediation is warranted. BLM reports that 40,652 of the 45,312 abandoned mine sites pose only physical safety hazards. However, 30,553 of these 40,652 physical safety sites need verification or initial assessment to determine whether environmental hazards are present.

The number of potentially contaminated mines may be larger than BLM's inventory indicates because BLM has not identified all of the abandoned mines on the land it manages. BLM estimates that there may be approximately 100,000 abandoned mines that have not yet been inventoried in three selected states, and that it would take 2 to 3 years to complete the estimates for the other nine BLM states. BLM estimates that it will take decades to complete the inventory. To inventory a site, BLM field staff must visit the site to collect data, research the land ownership and extent of mining activity that occurred, and record the information in BLM databases.

The reliability of BLM's inventory of abandoned mines has been a long-standing concern. In 2005, Interior's Office of Inspector General (OIG) reported that BLM's national inventory of abandoned mines was incomplete, inaccurate, and inconsistent.[24] In addition, the OIG found that the BLM field office staff responsible for abandoned mines was not identifying or entering known, high-priority abandoned mine sites into the inventory database. Many of the sites listed in the database were obtained from old Bureau of Mines data that was never verified by site visits. Given the limited funds available, the OIG noted that it was important that the inventory include current and credible information needed for program management of significant sites. On this basis, the OIG recommended that BLM validate existing inventory data and develop procedures for ongoing data collection to ensure that data in the inventory is complete, accurate, and consistent.

BLM agreed with the recommendation and has made an effort to improve data quality and the inventory's completeness. BLM has an ongoing effort to estimate the number of abandoned mines and mine features that have not yet been inventoried on BLM lands and the approximate cost to complete the inventory. BLM established inventory teams in several states to go out and identify sites. [25] In addition, BLM began an initiative in California to determine the number of sites that need to be inventoried after the state provided the agency with digitized maps of potential mine sites and verified a sample of the sites. For California, BLM estimated that 22,728 sites and 79,757 features need to be inventoried. Of these, BLM estimated that approximately 2,600 sites will have environmental hazards. BLM estimates that it will cost approximately $120 million to confirm the inventory of 22,728 sites. BLM has estimated approximately 69,000 and 4,000 sites remain to inventory in Nevada and Utah, respectively, on BLM land. BLM officials told us that they expect to provide a report to Congress on the inventory work remaining in these three states in 2015. The nine remaining states with BLM land do not have the digital geographic data available that BLM has been using for California, Nevada, and Utah, according to BLM officials, making it difficult for BLM to develop similar estimates for these states.[26] BLM officials told us that the U.S. Geological Survey is working on an effort to develop datasets similar to those used to estimate the number of abandoned mines on the BLM land in California, Nevada, and Utah.

In addition, BLM has provided a contractor approximately $1 million to assist with entering additional sites in its database from records stored in local BLM offices. The contractor has also been working to clean the database of duplicate records, ensuring location data is correct, adding names to unnamed records, and conducting field verification of sites. According to BLM, the inventory of abandoned mines in the database was 19,000 in 2008, and it has increased to over 45,000. Figure 1 shows the known BLM abandoned mine sites in the western United States.

After BLM, the NPS is the Interior agency with the largest number of potentially contaminated and contaminated sites. Like BLM, NPS has hazardous materials sites, such as landfills and underground storage tanks, and abandoned mine sites. NPS has identified 377 hazardous materials sites where the presence or likely presence of contamination is at concentrations significant enough to require further assessment or cleanup. NPS also has identified 96 locations of concern. These are sites where contamination is suspected based on known past activities or observed and reported physical indicators but which require further assessment. Nearly half of these are old dump sites.

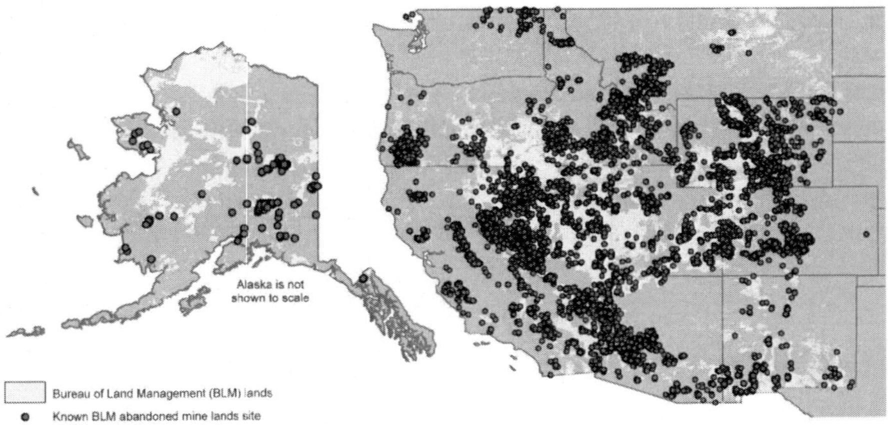

Source: BLM. | GAO-15-35.

Figure 1. Known BLM Abandoned Mine Sites in the Western United States.

In 2013, NPS completed a system-wide inventory and assessment project to identify abandoned mines on lands they manage and entered this information in their “Abandoned Mineral Lands Database.” NPS used specialists trained in natural sciences with knowledge of mining and exploration methods, equipment, and impacts to develop the inventory.[27] NPS’s inventory identified 37,050 mine features at 3,421 sites on NPS land. Of the total inventory, NPS believes that 3,841 features at 1,270 sites still require some level of effort to address human health and safety and/or environmental concerns. According to the NPS, there are 1,270 sites that require action, there are 227 sites in 76 parks that have or may have environmental contamination: 40 sites are confirmed, and 187 sites are locations of concern. According to NPS officials, these locations of concern require verification or initial assessment to determine if there are environmental hazards. According to the NPS, the remaining 1,043 abandoned mine sites pose physical safety hazards rather than environmental hazards. As a result of this effort, officials with NPS’s Abandoned Mineral Lands Program told us that they now believe that their inventory of all potentially contaminated sites is largely complete.

Bureau of Indian Affairs, Bureau of Reclamation, and Fish and Wildlife Service also have sites with environmental contamination. Officials from each of these agencies told us that they believe their inventories of sites with environmental contamination are complete. In addition to contaminated sites, the Fish and Wildlife Service has a number of potentially contaminated locations of concern at wildlife refuges and hatcheries that need to be investigated to determine if contamination exits. Table 2 below notes the number of contaminated and potentially contaminated sites for each of these agencies.

Table 2. Potentially Contaminated and Contaminated Sites for Selected Agencies

Agency	Locations of concern	Sites with known orsuspected contamination
Bureau of Indian Affairs	4	160
Bureau of Reclamation	[a]	15
Fish and Wildlife Service	394[b]	32

Sources: Locations of concern data was provided by each individual agency. The source of contamination known or suspected is from Interior's Environmental and Disposal Liability database as of the close of fiscal year 2013. | GAO-15-35

[a]The Bureau of Reclamation's inventory of locations of concern is maintained in its regional offices, and we did not collect this information because we determined that it was impractical to do so given the relatively small amount of land managed by this agency.

[b]The U.S. Fish and Wildlife Service's field stations (refuges and hatcheries) have identified locations of concern at 394 locations. This does not equate to 394 locations of concern; rather, it is the number of field stations identifying a location of concern.

USDA and Interior Use Funds from a Variety of Sources to Assess and Clean Up Sites

USDA and Interior use funding from annual appropriations to assess and to clean up sites. In addition, one of the USDA agencies uses funds it is authorized to borrow to carry out assessment and cleanup activities. USDA and Interior have also used one-time Recovery Act funds[28] for these purposes. In addition, both USDA and Interior seek cost recovery funds from potentially responsible parties to help offset cleanup costs at sites where those parties caused or contributed to contamination.

USDA

In fiscal year 2013, USDA allocated over $22 million to environmental cleanup efforts. Specifically, USDA allocated (1) $3.7 million for department-wide cleanup projects, the majority of which were for cleanup at USDA's Beltsville site and to cover legal expenses; (2) $4.3 million in funds to mitigate contamination at former grain storage sites; and (3) approximately $14 million for the Forest Service to conduct environmental assessments and cleanup activities.[29]

USDA's funds for environmental cleanup programs declined annually for fiscal years 2003 through 2013—with the exception of 2009, when USDA used Recovery Act funds for this purpose. USDA's funds over this period were composed of USDA's department-wide Hazardous Materials Management Appropriation (HMMA), the Forest Service's Environmental Compliance and Protection (ECAP) funds, funds the CCC is authorized to borrow, and one-time Recovery Act funds for the Forest Service.[30]

- HMMA is USDA's department-wide account that is available to pay for cleanup activities, among other things.[31] Of USDA's three annual sources of environmental cleanup funding, the HMMA account has experienced the most precipitous decline. According to USDA, HMMA funds declined from $15.7 million in fiscal year 2003 to $3.7 million in fiscal year 2013—a decrease of over 76 percent (see fig. 2). As a result, USDA has been able to fund fewer cleanup activities. In fiscal year 2003, USDA funded 110 cleanup and support activities at 81 sites and USDA offices with HMMA funds. In contrast, in fiscal year 2013, USDA was only able to fund 4 cleanup and support activities with HMMA funds. USDA officials said that these funds have decreased as a result of decreases in appropriations.

The ECAP program is the primary source of funding for the Forest Service's environmental cleanup programs. The Forest Service allocates approximately $500,000 a year to each of its nine regional offices for a variety of activities including preliminary assessments, site inspections, and small cleanups. The Forest Service awards the remaining funds to its regional offices for larger, high-priority assessments and cleanups on a priority basis. The ECAP program has also experienced a decline in recent years. USDA officials said that these funds have also decreased as a result of decreases in appropriations. ECAP funds have declined from $28 million in fiscal year

2003 to $14.4 million in fiscal year 2013—a decrease of 48 percent.[32] The ECAP program does not have historical records on ECAP projects going back to fiscal year 2003, so we were not able to compare the number and type of cleanup projects from that time with similar projects in recent years. However, in fiscal year 2013, the ECAP program funded 61 projects, including assessments, cleanup activities, PRP oversight, and maintenance activities. Figure 3 shows the decline in funds.

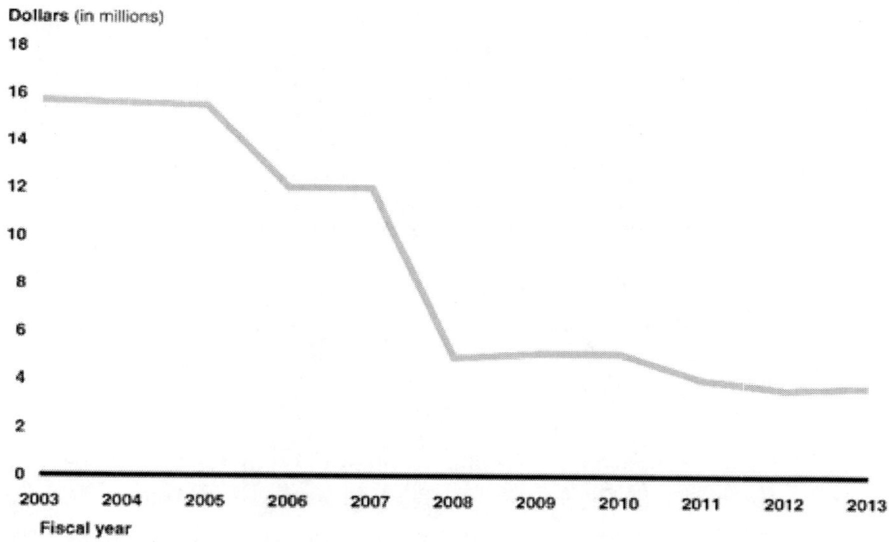

Source: USDA. | GAO-15-35.

Figure 2. USDA's Hazardous Materials Management Appropriations (HMMA), Fiscal Years 2003-2013.

- The CCC uses up to $5 million annually from funds it is authorized to borrow,[33] and it also requests HMMA or other funds when needed. CCC uses these funds to mitigate contamination at its former grain storage sites. CCC officials told us that contamination at the former grain storage sites tends to be widespread, making cleanup difficult, expensive, and time-consuming. For fiscal years 2003 through 2013, CCC assessed 31 sites, and it initiated or completed cleanup at 9 sites. Twelve sites also received No Further Action status.
- The Forest Service also allocated approximately $20 million in one-time Recovery Act funds in fiscal year 2009 to cleanup activities at 14 sites located on, or directly impacting, land managed by the Forest Service.

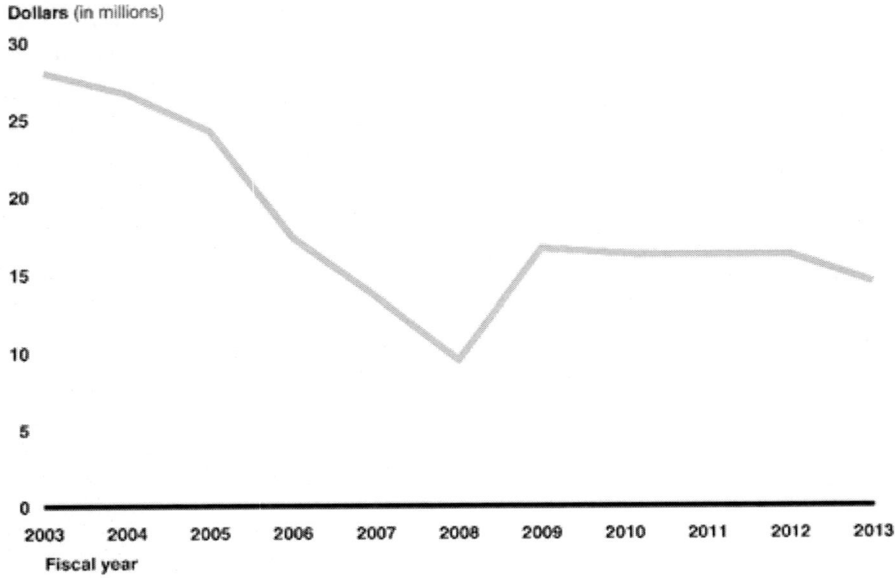

Source: USDA. | GAO-15-35.

Figure 3. Forest Service Environmental Compliance and Protection (ECAP) Funding Amounts, Fiscal Years 2003-2013.

In addition to these four funding sources, USDA seeks recovery of cleanup costs and natural resource damages under CERCLA from potentially responsible parties, such as owners and operators of a site, to help offset cleanup costs at sites where they caused or contributed to contamination.[34] Cost recovery amounts vary from year to year. Between fiscal years 2003 and 2013, USDA typically recovered $30 million or less. However, according to agency documents, USDA successfully recovered over $170 million from a single mining company as part of a bankruptcy case in 2009. These funds were used to conduct cleanup activities at 13 mine sites located on NFS lands. In fiscal year 2011, USDA recovered $65 million from another mining company for restoration of injured natural resources in the Coeur d'Alene River Basin NPL site in Idaho.

Currently, the majority of USDA's environmental cleanup funds are spent cleaning up ARS's Beltsville NPL facility and abandoned mines and landfills on NFS lands, as well as mitigating potential groundwater contamination from activities at former CCC former grain storage sites. In 2003, USDA established the CEE database to track progress on all hazardous waste cleanup projects funded with HMMA funds. However, a series of budget cuts led

USDA to scale back and repurpose the database. Currently, CEE serves to gather accomplishment data—such as site identification efforts, assessments, and cleanup activities—for the Forest Service, regardless of funding source. Because most of USDA's contaminated sites are located on land managed by the Forest Service, CEE data provide an important snapshot of USDA's historical accomplishments. A review of CEE data and Forest Service Region 10 data show that the Forest Service has conducted at least 714 assessments [35] and cleaned up 371 sites[36] between fiscal years 2003 and 2013. EMD issued an accomplishment report in 2011 and 2012 based on CEE data and supplemental agency data, such as data from its annual call letter. According to a USDA official, EMD plans to continue issuing such reports on an annual basis.[37]

Interior

In fiscal year 2013, Interior allocated about $13 million for environmental cleanup efforts. Specifically, Interior allocated $10 million for cleanup projects department-wide, NPS allocated an additional $2.7 million and the Fish and Wildlife Service allocated over $800,000 for environmental assessment and cleanup projects. In addition to these project funds, BLM allocated more than $34 million to its hazardous management and abandoned mine programs. BLM provided over $18 million of these funds to its state offices; however, the amount of these funds specifically used for environmental cleanup projects was not readily available.

From fiscal years 2003 through 2013, Interior allocated over $148 million in Central Hazardous Materials Fund (CHF) resources to its agencies to support response actions undertaken at contaminated sites under CERCLA. This amount includes over $49 million in CHF "cost recoveries"—that is, funds recovered from or paid in advance to Interior by potentially responsible parties for past or future cleanup costs. To be eligible for CHF funding under agency policy, a site must have a completed preliminary assessment/site inspection, pose a significant risk to human health or the environment, and have a potentially responsible party search initiated or completed, among other things. Interior's agencies nominate projects for CHF funding. A technical review committee evaluates the project nominations and recommendations are forwarded to the Deputy Assistant Secretary for Policy and International Affairs for approval. Interior's agencies undertook 101 projects with CHF funding during fiscal years 2003 through 2013. These projects support a range of activities, from project oversight to advanced studies (e.g., remedial investigations, feasibility studies, engineering evaluations, and cost analyses) to removal and remedial actions. The majority of sites receiving CHF funding were abandoned mines, landfills, and former industrial facilities. In fiscal year 2013, Interior allocated $10 million to the CHF.

As previously noted, within Interior, BLM has the largest inventory of sites with potential environmental contamination and contaminated sites. BLM has two programs that address environmental contamination, among other things. The Abandoned Mine Lands Program works to address both environmental contamination and physical safety issues at mines. The Hazardous Materials Management and Resource Restoration Program addresses environmental issues across all BLM-managed lands, and it may help address environmental issues at high-priority abandoned mine sites. In fiscal year 2013, BLM allocated over $34 million to these two programs. Specifically, in fiscal year 2013, BLM allocated $16 million to the Hazardous Management and Resource Restoration Program. Funding for the program has fluctuated between approximately $16 million and $17 million each fiscal year since 2003. Funding for the Abandoned Mine Lands Program increased by almost $6.5 million in fiscal year 2009, and the program has retained these increased funds. In fiscal year 2013, BLM allocated $18.7 million to the Abandoned Mine Lands Program.

Of the $18.7 million funds allocated in fiscal year 2013, BLM provided over $4.5 million to BLM state offices for hazardous waste projects at mines.[38] Using funds received from both the Hazardous Materials Management and Resource Restoration Program and Abandoned Mine Lands Program, BLM provided over $14 million to the BLM state offices to cover program operations, expenses, and projects. According to BLM officials, after the BLM state offices fund program operations and expenses, they spend the remainder of the funds on (1) physical safety projects at abandoned mines and (2) environmental projects at abandoned mines or other sites needing cleanup. However, BLM headquarters does not compile information on the amount of funds each BLM state office had for such projects, the number of projects state offices conducted, or the amount spent on projects. As such, headquarters officials could not provide us with information on the amount of funds spent on environmental projects versus physical safety projects.

BLM also spent over $27 million in one-time Recovery Act funds on physical safety and/or environmental remediation projects at 76 locations. According to BLM, there were 31 projects for environmental activities.

BLM officials told us that the current funding levels are not sufficient to complete the inventory and address the physical and environmental hazards at abandoned mines. In its 2014 and 2015 budget justifications, Interior described proposals to charge the hardrock mining industry fees and use the funds to address abandoned mines. Interior proposes to levy a fee on all uranium and metallic mines on both public and private lands. The proposed fee would be charged per volume of material displaced. The fees collected would be distributed to two sets of funds, one for federal lands and another for nonfederal lands. The Secretary would disperse the share of non-federal funds to each state and tribe based on need. Each state and tribe would select its own priority projects using established criteria.

NPS allocates an average of $2 to $3 million each year to its own cleanup fund, and it allocated $2.7 million in fiscal year 2013. NPS uses a combination of risk factors to rank sites for receiving these funds, including human risk, ecological risk, and the potential for leveraging funds from other responsible parties. An NPS official told us that the agency has inadequate funding to address its over 400 potentially contaminated and contaminated sites. The agency has been able to address its highest risk sites with funding from the CHF. If there is a very significant risk, NPS can usually obtain funds to address the portion of the site that has the highest risk, if not the site as a whole. According to NPS officials, NPS has not selected response actions for almost 300 sites because current funding levels are not sufficient to address them.

The Fish and Wildlife Service uses project funds from its Refuge Cleanup Fund. The National Wildlife Refuge System set aside $1.7 million in annual appropriations in fiscal year 2003, but these funds have decreased over the last decade to $834,000 in fiscal year 2013. The Fish and Wildlife Service uses a portion of these funds for environmental compliance work. For example, in fiscal year 2013, the Fish and Wildlife Service approved funding of about $700,000 for 11 projects such as an underground storage tank remediation and a site investigation at a firing range. Fish and Wildlife Service officials review proposals for cleanup projects on national wildlife refuges and select sites for the funds using a ranking system based on risk to human health and the environment, legal requirements, and regional priorities. For emergency response to chemical and hazardous material accidents or immediate removal, Fish and Wildlife officials said the agency can use funds from an emergency construction account.

Interior tracks the progress of cleanup at sites listed as environmental and disposal liabilities in the Environmental Management Information System on a quarterly basis. The stage of cleanup for 756 sites is provided in appendix III.

USDA AND INTERIOR REPORT OVER $350 MILLION IN ENVIRONMENTAL LIABILITIES IN THEIR ANNUAL FINANCIAL STATEMENTS

USDA and Interior reported environmental liabilities of $176 million and $192 million, respectively, in their fiscal year 2013 financial statements. These liabilities represent what the agencies have determined to be the probable and reasonably estimable future costs to address 100 USDA sites and 434 Interior sites, as required by federal accounting standards.

In its fiscal year 2013 financial statements, USDA reported a total of $176 million in environmental liabilities. The $176 million amount included: $165 million to address asbestos contamination,[39] $8 million for up to 76 CCC former grain storage sites in the Midwest that are contaminated with carbon tetrachloride,[40] and $3 million for 24 Forest Service sites, including guard stations, work centers, and warehouses, among others.[41] USDA also reported $40 million in contingent liabilities[42] for environmental cleanup at four phosphate mine sites in Southeast Idaho.[43]

Interior reported $192 million in environmental liabilities in its fiscal year 2013 financial statements. This number reflects the low end of a range of probable future costs for completing cleanup activities at 434 sites. These activities include studies or removal and remedial actions at sites where Interior has already conducted an environmental assessment and where Interior caused or contributed to the contamination or has recognized its legal obligation for addressing the site. Interior also disclosed in notes to its financial statements the estimated cost range for completing cleanup activities at these sites. The cost range disclosed was approximately $192 million to $1.3 billion. The difference between the upper and lower ends of the range ($1.1 billion) reflects the additional potential cost for addressing these sites.

Interior also disclosed the estimated costs for government acknowledged sites where it was reasonably probable that cleanup costs would be incurred. In fiscal year 2013, Interior disclosed in the notes to its fiscal year 2013 financial statements a cost range for these activities to be approximately $62 million to $139 million.[44] The majority of this cost range was related to addressing 85 abandoned mine sites.

As USDA and Interior complete their inventories and assess them for contamination, it might be expected that related environmental costs for both agencies will increase as additional sites are considered.

EPA's Primary Actions with Regard to Interior and USDA Sites Are Maintaining a List of Potentially Hazardous Sites and Overseeing the Cleanup at One NPL Site

EPA's primary roles with regard to contaminated and potentially contaminated sites under USDA's or Interior's jurisdiction include (1) placing the sites on the docket, (2) taking steps to assure that assessments of the sites are conducted, (3) determining whether the sites should be proposed for listing on the NPL, and (4) overseeing the cleanup of sites placed on the NPL. From information provided by the federal agencies, EPA has compiled a docket of federal sites that may pose a risk to human health and the environment, including 260 USDA sites and 528 Interior sites. Furthermore, EPA continues to review new information provided by USDA and Interior to determine if any additional sites should be added to the docket. EPA also is responsible for ensuring that USDA and Interior assess these sites for contamination; USDA and Interior have assessed the majority of these sites. EPA has also placed a number of USDA and Interior sites on the NPL and is currently overseeing the ongoing cleanup of one USDA NPL site.

EPA Is Responsible for Placing Certain Sites on the Federal Agency Hazardous Waste Compliance Docket

EPA is responsible for compiling a list of potentially contaminated federal facilities reported by agencies under certain provisions of RCRA or CERCLA. This list is the Federal Agency Hazardous Waste Compliance Docket (docket). EPA compiled the first docket in 1988 and CERCLA requires EPA to publish a list of any new sites added to the docket in the *Federal Register* every 6 months. EPA updates the docket after receiving and reviewing notices from federal agencies concerning the generation, transportation, treatment, storage, or disposal of hazardous wastes or release of hazardous substances. Specifically, section 120(c) of CERCLA requires the agency to update the docket with information submitted by federal agencies under the following statutory provisions:

- Section 3005 of RCRA, which requires owners and operators of facilities where hazardous waste is treated, stored, or disposed ("TSD facilities") to provide certain information to EPA when applying for permits.

- Section 3010 of RCRA, which requires owners and operators of TSD facilities, as well as generators and transporters of hazardous waste, to notify EPA of their hazardous waste activities.
- Section 3016 of RCRA, which requires federal agencies to prepare a biennial inventory of hazardous waste sites currently or previously owned or operated.
- Section 103(a) of CERCLA, which requires facility owners/operators to notify the National Response Center of any reportable releases of hazardous substances.

EPA officials told us if agencies fail to comply with hazardous waste self-reporting requirements and do not report a site, it is difficult for EPA to know about the site to list it. However, if EPA learns about a site that has had a release or threat of a release of hazardous substances through other means, EPA will list the site on the docket.

It is EPA's policy to exclude certain categories of sites from docket listing. For instance, small quantity generators of hazardous waste (less than 1,000 kilograms in any month) are excluded from docket listing unless they have reported releases.[45] While EPA lists some abandoned mines on the docket, in 2003, EPA issued a policy statement that mines with "mixed ownership"— mines located partially on private land and partially on public land — should typically not be included on the docket. Instead, individual mines should be evaluated on a case by case basis. Under this approach, EPA has listed a number of mixed ownership mines on the docket.

EPA has published 26 updates of the docket as of September 2014, but the agency has not consistently met the 6-month reporting requirement. Agency officials said that they have not met this requirement in recent years due to staff efforts to include as many facilities as possible in each docket publication by enabling longer quality assurance reviews of facilities that were identified as potential docket sites. Prior to 2014, the effort to compile and monitor the docket listings was a manual process. However, in 2014, EPA implemented revised docket procedures with a computer based process that is to compile potential docket listings from agency notices by searching electronic records. According to EPA officials, the first docket update with the new system will be published by the end of fiscal year 2014. EPA officials said that they expect the new system to allow them to update the docket in a more timely way in the future.

Since 1988, EPA has listed 260 USDA sites and 528 Interior sites on the docket. However, EPA, Interior, and USDA officials disagree over whether some of these sites should have been listed. EPA and Interior officials disagree about the overall approach of listing sites on the docket that are not reported to EPA by an agency. Interior officials believe that CERCLA does not give EPA the discretion to list sites unless Interior reports the sites to EPA and that EPA should limit its listing of sites on the docket to those reported by an agency under one of the provisions specifically noted in CERCLA. Interior and USDA officials also believe that abandoned mines should not be listed on EPA's docket because the agencies did not cause the contamination, and, therefore, the sites should not be considered federal sites. However, EPA officials believe that, regardless of whether USDA and Interior are legally liable for addressing these sites, they have an independent responsibility under Executive Order 12580[46] and CERCLA as land management agencies owning the sites to address them.

EPA's docket and USDA's and Interior's inventory lists differ in a number of ways. First, the docket includes a historical record of potentially contaminated sites that typically have been reported to EPA by the agencies. Because it includes a historical record, sites that subsequently were found to not be contaminated and sites that the agencies may have addressed are still included on the docket. Furthermore, agency inventories may include potentially contaminated sites that the agency is not required to report to EPA for inclusion on the docket. For example, an abandoned mine that is not known to have had a reportable release of a hazardous substance may not have to be reported to EPA.

EPA Is Responsible for Taking Steps to Assure That Preliminary Assessments Are Conducted for Docket Sites

After a site is listed on the docket, CERCLA requires EPA to take steps to assure that a preliminary assessment is conducted. EPA has established 18 months as a reasonable time frame for agencies to complete the preliminary assessment. However, EPA cannot enforce the requirement or penalize agencies for not completing a preliminary assessment within the 18-month time frame. As of February 2014, both Interior and USDA had conducted a preliminary assessment of the majority of their sites on EPA's docket. However, EPA, Interior, and USDA have differing information on the status of preliminary assessments for their docket sites.

Our analysis of data in EPA's Comprehensive Environmental Response, Compensation, and Liability Information System showed that USDA had completed preliminary assessments for 186 of the 260 USDA sites on the docket and submitted a preliminary assessment for 24 sites where EPA has yet to make a decision about whether the preliminary assessment is complete. According to EPA's data, USDA still needs to conduct a preliminary assessment at 50 docket sites. We reviewed the 50 sites with USDA officials. These officials told us that they believe no further action is required at 48 of these sites. They noted that preliminary assessments had been conducted at 34 of the sites, including locations where USDA's records indicate that a preliminary assessment was provided to EPA, cleanup actions were completed, or USDA had notifications from EPA or state environmental agencies that no further action is required. Of the remaining 16 sites, USDA officials said that a time critical removal of hazardous waste was conducted at 4 sites in lieu of a preliminary assessment. USDA officials said that they had no plans to conduct a preliminary assessment at 8 of the sites because a hazardous release had not occurred. These sites are listed on the docket because of their past or present operation as a disposal facility—not because of any identified release of hazardous waste—that triggered a notice to EPA.[47] According to USDA, these sites are listed on the docket because of their past or present operation as a one time or periodic large quantity RCRA generator, not because of any identified release of hazardous waste. USDA said that it does not believe that facilities in good standing that handle hazardous waste and generate a manifest under RCRA should then have to conduct a preliminary assessment as a result of a docket listing. A preliminary assessment is to be conducted for sites listed on the docket, however.[48] USDA believes that two sites are listed in error—one does not belong to USDA, and the other USDA believes is a duplicate listing. Table 3 shows the preliminary assessment status USDA provided on these 50 sites.

Based on our analysis of EPA data, Interior has conducted preliminary assessments at 442 of 528 docket sites. In addition, EPA has received 7 preliminary assessments from Interior that they are still reviewing. According to EPA's data, Interior still needs to conduct a preliminary assessment at 79 docket sites. We reviewed the status of the remaining 79 docket sites with Interior officials. These officials told us they believe they have met the preliminary assessment requirement at 47 of the 79 sites. According to Interior officials, these include sites where remedial investigations have been conducted, sites where cleanup activities are ongoing, and sites where cleanup activities are completed. In addition, the officials said that 1 site had a preliminary assessment in progress, and 3 of the docket sites still need a preliminary assessment. Interior officials do not have plans to conduct a preliminary assessment at 26 sites for a number of reasons. Interior reported that cleanup at 13 of the sites was the responsibility of another agency. According to Interior, three sites are small quantity generators, and the agency believes the sites should have been excluded from the docket. Interior does not have plans to conduct a preliminary assessment at 4 sites because they disagree with the listing. For some of these 4 sites, NPS officials said that they had no knowledge of a release at the site and that the agency had not provided any type of notice to EPA regarding the site. Table 4 shows the preliminary assessment status of the 79 docket sites as reported by Interior.

Table 3. USDA's Reported Preliminary Assessment Status of Disputed Docket Sites

Status	Number of sites
Preliminary assessment complete	34
No release occurreda	8
Time critical removal in lieu of preliminary assessment	4
Undetermined	2
Otherb	2
Total	50

Sources: USDA and GAO analysis of EPA data. | GAO-15-35.

[a]These sites are listed on the docket because of their past or present operation as a disposal facility, not necessarily because of any identified release of hazardous waste.

[b]USDA believes one of these sites is the responsibility of another agency and that one is a duplicate docket listing.

According to EPA officials, USDA and Interior have different opinions on the status of a preliminary assessment for a number of reasons. First, USDA and Interior may have conducted preliminary assessment work and submitted information to EPA, but EPA officials may not have concurred that the work was sufficient to meet the preliminary assessment requirement. Second, the agencies may not have submitted information to EPA about the preliminary assessment at the site. Third, while EPA carefully considers USDA and Interior comments regarding proposed docket sites before listing, USDA and Interior do not always agree with final docket listings. USDA and Interior generally do not perform a preliminary assessment if they do not agree with

EPA's determination that a facility should have been added to the docket. For example, EPA officials said that, even if an agency removes hazardous waste from a site, the site may still meet the legal requirements for being added to the docket, and a preliminary assessment at the site may still be needed to determine if there is additional contamination.

Table 4. Interior Reported Preliminary Assessment Status of Disputed Docket Sites

Status	Number of sites
Preliminary assessment conducted	47
Responsibility of DOD or U.S. Army Corps of Engineers	13
Unknown sitea	6
Disagreement with docket listing	4
Preliminary assessment needed	3
Small quantity generator	3
Othera	2
Preliminary assessment in progress	1
Total	79

Sources: Interior and GAO analysis of EPA data. | GAO-15-35.

[a]Unknown sites are sites listed on the docket that did not list a responsible agency name. We sent a list of these sites to the Interior agencies and no agency recognized or claimed these sites. Other includes two docket sites for the U.S. Geological Survey that we did not include in the scope of this review.

EPA officials said that they have an ongoing initiative to verify the status of federal sites in EPA's database. In EPA's 2014 and 2015 budget request, EPA stated that it will begin working cooperatively with states and federal facilities on a multiyear effort to complete outstanding facility assessments and close the compliance gap. However, EPA officials with the Federal Facilities and Restoration and Reuse Office told us that they received 20 percent less funds in fiscal year 2014 and that this may negatively affect their completion of this effort.

Without accurate and consistent information from USDA and Interior on the status of their sites, EPA cannot determine whether additional work is needed to complete the preliminary assessments required by CERCLA and update its records. Furthermore, unless EPA informs USDA and Interior of its determination that preliminary assessments are not complete, USDA and Interior will not be aware of EPA's expectations that additional work is necessary for the preliminary assessments to be complete. As a result, EPA cannot effectively assure that preliminary assessments are being conducted in a timely manner and, ultimately, that sites most in need of remediation are addressed.

EPA Is Responsible for Evaluating Whether Federal Sites Should Be Placed on the National Priorities List

After the agency conducts the preliminary assessment, EPA reviews it to determine whether the information is sufficient to assess the likelihood of a hazardous substance release, a contamination pathway, and potential receptors. EPA may determine that the site does not pose a significant threat and requires no further action.[49]

If further investigation is needed, EPA may request that the agency conduct a site inspection to gather more detailed information. If on the basis of the site inspection, EPA determines that hazardous substances, pollutants, or contaminants have been released at the site, EPA will use the information from the preliminary assessment and site inspection to calculate and document a site's preliminary Hazard Ranking System (HRS) score, which indicates a site's relative threat to human health and the environment based on potential pathways of contamination. The HRS scores sites on four possible pathways: groundwater migration, surface water migration, soil exposure, and air migration. Sites with an HRS score of 28.50 or greater become eligible for listing on the NPL. According to EPA, in order to maintain close coordination with the states and tribes in the NPL listing decision process, EPA's policy since 1996 has been to determine the position of the states and tribes on sites that EPA is considering for listing.

To list a site on the NPL, EPA must publish a proposed regulation and give the public notice and an opportunity for comment before finalizing it. According to EPA officials, EPA and OMB have determined that the NPL rulemakings do not fall under Executive Order 12866 requiring review by OMB.[50] However, EPA informally submits both NPL proposed and final rulemaking packages to OMB for review. During its review process, OMB will discuss the package with the federal agency that manages the site and other federal agencies. Some proposed sites do not proceed past this step and do not become NPL sites.

After OMB concurs with the listing package, EPA publishes a notice of proposed rulemaking in the *Federal Register*, obtains public comments during the specified period, reviews the comments received, responds to comments, and makes a final listing decision. The final listing decision process includes input from states and tribes and concurrence from OMB. EPA publishes the final rule in the *Federal Register*, formally adding the site to the NPL.

According to EPA, the majority of USDA and Interior sites on the docket have not required further action from EPA after review of the preliminary assessment. However, a number of sites have been eligible for listing on the NPL. As of September 2014, two USDA sites and two Interior sites were on the NPL. In 2009, EPA considered proposing one Interior site, Red Devil Mine in Alaska, but it was ultimately never proposed for NPL listing. BLM objected to the listing, arguing that the mercury mine was abandoned before federal land managers had any practical regulatory authority over this type of site,[51] and that listing these sites as "federal facilities" on the NPL would impede BLM's efforts to hold private parties responsible for cleanup at similar mine sites. As an alternative to listing, EPA and BLM entered into negotiations for an agreement that would clarify the roles and responsibilities of BLM, EPA, and the state of Alaska. According to EPA officials, EPA's goal is to establish a framework for cleanup that satisfies CERCLA regulatory requirements. Specifically, EPA wants the agreement to provide for (1) public participation and community involvement; (2) meaningful participation by local tribes, including appropriate opportunities for tribal consultation; (3) EPA concurrence on cleanup choices, in consultation with the state, as well as concurrence on work plans, reports, and other decision documents; (4) enforceable milestones; and (5) a formal dispute resolution process in which all parties accept EPA's final decision. BLM objects to the enforceability and dispute resolution provisions. For this and other reasons, since 2009, BLM and EPA have not been able to reach an agreement. The state of Alaska asked EPA to list the site on the NPL in 2012 and 2013. BLM continues to work at the site and informally coordinate with EPA and the Alaska Department of Environmental Conservation.

EPA Oversees Cleanup of NPL Sites, and Work at Most USDA and Interior NPL Sites Is Complete

Once a site is on the NPL, EPA oversees the cleanup. As part of its oversight responsibility, EPA works with the agency to evaluate the nature and extent of contamination at a site. The federal agency must then enter into an interagency agreement with EPA that includes: (1) a review of remedial alternatives and the selection of the remedy; (2) schedules for completion of each remedy; and (3) arrangements for the long-term operation and maintenance of the site. According to EPA, the agreements also provide a process for EPA and the federal agency to resolve any disagreements related to implementing the cleanup remedy, with EPA being the final arbiter of such disputes. Once the agency and EPA agree on a cleanup remedy, the agency implements the remedy at the site. Afterwards, the agency must conduct long-term monitoring to ensure the remedy remains protective of human health and the environment. For sites where hazardous substances, pollutants, or contaminants were left in place above levels that do not allow for unlimited use and unrestricted exposure, every 5 years following the initiation of the remedy, the agency must conduct a formal review of the site and provide it to EPA. EPA determines whether the selected remedy is still protective of human health and the environment.

For three of the four USDA and Interior sites on the NPL, cleanup remedies have been selected, and construction completed or implemented. EPA has conducted at least one 5-year review at each of

these three sites and determined that the site's selected cleanup remedy is still protective of human health and the environment. Of the four NPL sites, only USDA's Beltsville Agricultural Research Center site is undergoing active cleanup. Appendix IV provides additional information on these four NPL sites.

Conclusion

Interior and, to a lesser extent, USDA have made progress in identifying contaminated and potentially contaminated sites on lands they manage. However, neither agency has completed the task, primarily because of the significant effort required to locate abandoned mines on their vast acreages. Interior agencies have developed separate national inventories of contaminated and potentially contaminated sites, and BLM continues to take steps to expand the number of sites in its abandoned mine database and improve the accuracy of its data. Continuing progress in this effort should help Interior develop a more comprehensive inventory that it needs to most effectively manage its cleanup programs. USDA, on the other hand, has no reliable national inventory of potentially contaminated sites or plans and procedures for developing one. In particular, USDA's data on the potentially contaminated abandoned mines on its lands is unreliable. Without a comprehensive, reliable, and easily accessible inventory of potentially contaminated sites on lands managed by USDA, including abandoned mines, and plans and procedures for developing such an inventory, USDA cannot either effectively manage the department's cleanup programs or ensure that limited funds are targeted to clean up those sites that may pose the greatest threats to human health or the environment. However, given USDA's decreasing funding for environmental restoration programs, current funding may be inadequate to both complete this inventory effort and continue cleanup efforts on sites with known contamination.

Under CERCLA, EPA is responsible for assuring that preliminary assessments are conducted at agency sites listed on the docket. However, EPA, USDA, and Interior disagree over how many docket sites need a preliminary assessment, and the agencies have differing information on whether this requirement has been met at a number of sites. Without accurate and consistent information from USDA and Interior, EPA cannot resolve whether additional work is needed and update its Superfund Enterprise Management System database. More importantly, without this information, EPA cannot effectively assure that preliminary assessments are being conducted in a timely manner and that sites most in need of remediation will be addressed.

Recommendations for Executive Action

To ensure that USDA has the information needed to better identify potentially contaminated sites—particularly abandoned mines—on properties it manages and, thereby, help minimize possible risks to human health and the environment, GAO recommends that the Secretary of Agriculture direct the heads of the department's land management agencies to develop plans and procedures for completing their inventories of potentially contaminated sites.

To help resolve disagreements between EPA and USDA and Interior regarding which remaining docket sites require preliminary assessments, GAO recommends that the Administrator of EPA direct the Office of Federal Facilities Restoration and Reuse to take the following three actions:

- review available information on USDA and Interior sites where EPA's Superfund Enterprise Management System indicates that a preliminary assessment has not occurred to determine the accuracy of this information, and update the information, as needed;
- after completing this review, inform USDA and Interior whether the requirement to conduct a preliminary assessment at the identified sites

has been met or if additional work is needed to meet this requirement; and

- work with the relevant USDA and Interior offices to obtain any additional information needed to assist EPA in determining the accuracy of the agency's data on the status of preliminary assessments for these sites.

AGENCY COMMENTS AND OUR EVALUATION

We provided a draft of this report to Interior, USDA, and EPA for review and comment. Interior, USDA, and EPA also provided technical comments, which we incorporated, as appropriate.

In an e-mail we received December 1, 2014, Interior's Audit Liaison stated that Interior generally concurred with our findings.

USDA submitted written comments, included in appendix V, stating that the department did not agree with the key findings in our draft report. With regard to our finding that USDA does not have a complete, consistent, or usable centralized inventory of abandoned mines and has no plans or procedures for developing one, USDA stated that it does have a centralized inventory and that this inventory is in a transition phase as a result of reduced funding levels. USDA also stated that it has taken a number of actions to manage its inventory in a more cost-effective manner, reduce operating costs, and eliminate data collection redundancies across USDA agencies.

Although USDA asserts that it has a centralized site inventory, USDA could not provide us with up-to-date information on sites across the department. Instead we had to obtain information from USDA's individual agencies. Furthermore, USDA stated that it relies on data from outside sources, such as states and other federal agencies, to supplement this information. However, because these data are from a variety of sources and not part of a centralized USDA database, USDA may not be able to ensure the quality, completeness, consistency, or reliability of this information.

USDA acknowledged that abandoned mine lands make up the bulk of its inventory of contaminated and potentially contaminated sites. USDA also recognized that its inventory of potentially contaminated sites is not complete. USDA stated that, given limited funding resources, it is utilizing abandoned mine data from all existing sources and developing collaborative partnerships with federal and state agencies. USDA strongly disagreed that an incomplete inventory leads to ineffective management of the program.

We continue to believe that USDA's inventory of contaminated and potentially contaminated sites—in particular, abandoned mines—is insufficient for effectively managing USDA's overall cleanup program. In our report, we note that USDA has a list of contaminated sites, but its inventory of potentially contaminated sites is incomplete and unreliable. For example, the Commodity Credit Corporation's list of potentially contaminated sites is incomplete because many states have not yet reported whether there is suspected contamination at former grain storage sites. In addition, USDA officials told us that because the data in both USDA's department-wide CEE database, and its AML database maintained by the Forest Service, are incomplete and contain inaccurate information, USDA could not provide a reliable number of the potentially contaminated sites the agency has identified. On this basis, we concluded that USDA does not have a complete, consistent, or usable inventory of potentially contaminated sites, in particular, abandoned mines. We continue to believe that, without reliable data, USDA cannot effectively manage its cleanup program or ensure that its limited funds are targeted to clean up sites that may pose the greatest threats to human health or the environment.

In its letter, USDA also expanded on its position with regard to its continuing disagreements with EPA over the need to assess certain USDA sites on the federal facilities docket. To the extent that the information USDA provided helps to clarify the department's position on this issue, we incorporated it into the report, as appropriate. Specifically, we added USDA's explanation of why it does not perform preliminary assessments if it does not agree with EPA's determination that these assessments are needed. In its letter, USDA stated that this situation generally relates to sites listed on the docket pursuant to RCRA. USDA stated that these sites are listed on the docket because of their past or present operation as a one time or periodic large quantity RCRA generator, not because of any identified release of hazardous waste. USDA said that it does not believe that facilities in good standing that handle hazardous waste and generate a manifest under RCRA should then have to conduct a preliminary assessment as a result of a docket listing. A preliminary assessment is to be conducted for all sites listed on the docket, however.

In written comments, reproduced in appendix VI, EPA generally agreed with our findings, conclusions, and recommendations. With regard to our recommendation that EPA review the status of USDA and Interior sites on the docket in order to identify whether a preliminary assessment is still needed, EPA stated that it will compare the site data provided by USDA and Interior during the review of existing Superfund Enterprise Management System data in order to identify sites that still need a preliminary assessment. With regard to our recommendation that EPA inform USDA and Interior whether the requirement to conduct a preliminary assessment has been met, or if additional work is needed, EPA stated that it will use the data collected from USDA and Interior to inform these agencies of their statutory requirements to complete site assessment work. Finally, as we recommended, EPA agreed to work with USDA and Interior to obtain additional information to determine the accuracy of EPA's data on the preliminary assessments for sites on the federal facilities docket. EPA stated that it believes that our recommendations will help to address the 50 USDA sites and 79 Interior sites with a preliminary assessment status that is being disputed; moreover, additional actions may be needed for other contaminated or potentially contaminated sites on USDA and Interior lands because EPA says it cannot compel agencies to comply with requests for preliminary assessments to be completed.

Sincerely yours,
J. Alfredo Gómez
Director, Natural Resources and Environment

APPENDIX I: OBJECTIVES, SCOPE, AND METHODOLOGY

This appendix details the methods we used to examine the Department of Agriculture's (USDA) and the Department of Interior's (Interior) efforts to identify and to clean up hazardous waste sites. We were asked to review the status of USDA's and Interior's efforts in this area. Specifically, this report examines: (1) the status of efforts to develop an inventory of potentially contaminated and contaminated sites, (2) the funding levels for addressing these contaminated sites, (3) the amount of environmental liabilities reported on agency financial statements, and (4) EPA's role in cleaning up these sites.

This review addresses the activities of five agencies within USDA: the Agriculture Research Service (ARS), Animal and Plant Health Inspection Service (APHIS), Commodity Credit Corporation (CCC), the Farm Service Agency (FSA), and the Forest Service. Within Interior, this review addresses the activities of five agencies: the Bureau of Indian Affairs (BIA), Bureau of Land Management (BLM), Bureau of Reclamation, Fish and Wildlife Service, and the National Park Service (NPS). We selected these agencies because they manage USDA and Interior lands.

Developing Inventories

To describe the status of efforts to develop an inventory of potentially contaminated and contaminated sites and determine the number of sites in each agency's inventory, we identified and analyzed current legal requirements for federal identification and reporting of potential hazardous waste sites, including the Comprehensive Environmental Response, Compensation, and Liability Act (CERCLA) and the Resource Conservation and Recovery Act (RCRA). We also obtained and reviewed USDA and Interior policies on identifying potentially hazardous waste sites. We interviewed agency officials from each agency identified above regarding efforts to identify potentially contaminated or contaminated sites, and the extent to which their inventories are complete. We reviewed relevant inventory data ranging from the period of September 2013 to June 2014 from 10 agency sources. Agencies with small inventories provided site numbers or a list. For the other agencies with more extensive inventories, we obtained information from a number of databases. Because the work to identify sites is conducted in the field offices of most of these 10 agencies, it was not practical for us to determine the extent to which agency field staff are following policies and procedures.

In addition, we obtained information from the agencies that ranged from September 2013 to June 2014 on the number of sites in each agency's inventory that the agency had identified as potentially contaminated but may not have visited yet to determine whether contamination exists, as well as the number of contaminated sites. We included sites that the agencies have identified as being on land managed or leased by USDA and Interior, including grain storage sites formerly leased by the CCC. We included CCC grain storage sites in our scope because USDA is currently expending funds to clean up these sites and has reported them as environmental liabilities on their annual financial statements.

USDA

To determine the number of contaminated and potentially contaminated sites for USDA's ARS, we obtained a list of all ARS facilities as of November 2013 and discussed the list with ARS officials. Because the list did not indicate which of ARS's facilities were contaminated or potentially contaminated, we relied on ARS officials to identify those facilities with known or possible contamination. ARS officials identified three contaminated and three potentially contaminated sites. Similarly, for APHIS, we relied on testimonial evidence from APHIS officials. According to APHIS officials, the agency does not own or operate many sites that deal with hazardous waste, so APHIS does not track its sites in a database or spreadsheet. However, based on their knowledge of APHIS sites, they identified two sites with known contamination and none that are potentially contaminated. For FSA and CCC (which is located within FSA), we met with an official representing FSA and CCC and reviewed the agency's Environmental and Disposal Liability spreadsheet, which tracks their inventory of former grain storage sites. Because there are only three foreclosure properties remaining in FSA's inventory, however, they do not track their foreclosure sites in a database or spreadsheet. We requested that FSA/CCC officials identify the former grain storage sites and foreclosure properties in their inventory that had confirmed or possible contamination.

Finally, for the Forest Service, we met with officials within the agency's headquarters office and all nine regional offices, as well as officials from USDA's EMD. Because Forest Service does not maintain an inventory of contaminated sites at the headquarters level, we requested and obtained data from EMD's Center for Environmental Excellence (CEE) database— which contains the agency's most current data on Forest Service sites in Regions 1 through 9. The CEE database does not track all of the agency's docket sites, however, or have current inventory data on contaminated sites in Forest Service Region 10. We, therefore, supplemented the CEE inventory data with data from EMD's Docket Status Spreadsheet and an inventory spreadsheet from Region 10 to calculate the Forest Service's total inventory of contaminated sites. To evaluate the reliability of these data and determine its limitations, we reviewed the data obtained from EMD and Forest Service Region 10, and we interviewed EMD and Forest Service officials. We analyzed related documentation, examined the data to identify obvious errors or inconsistencies and compared the data we received with other published data sources, such as USDA's annual environmental restoration accomplishment report. On the basis of our evaluation of these data, we concluded that the CEE data, Docket Status spreadsheet, and Forest Service Region 10 data we collected and analyzed were sufficiently reliable for the purpose of compiling a complete and accurate inventory of the Forest Service's contaminated sites. We also attempted to obtain Forest Service's inventory of potentially contaminated sites and requested information from the Forest Service regional offices, as well as data from CEE and USDA's Abandoned Mine Lands database. However, for the reasons described earlier in the report, we did not find the data to be reliable for purposes of calculating the Forest Service's inventory of potentially contaminated sites and, therefore, did not use the data for our report.

Interior

At Interior, we obtained the number of contaminated sites for BIA, the Bureau of Reclamation, the Fish and Wildlife Service, and NPS's hazardous materials sites from Interior's Environmental and Disposal Liability (EDL) database. These data were current as of September 30, 2013, and were the most recent data available at the time they were collected. We determined the data were sufficiently reliable for the purpose of reporting the agencies' 4,722 contaminated and potentially contaminated sites based on our review of Interior and agency handbooks and interviews with agency officials.

For the number of contaminated abandoned mines and potentially contaminated abandoned mines, we obtained data from NPS's Abandoned Mine and Site Cleanup Module database. We interviewed agency officials on the development of the database and reviewed the data entry instructions. We determined the data were sufficiently reliable for representing the 40 contaminated abandoned mines in NPS's abandoned mine inventory. To identify the number of sites in BLM's inventory, we obtained and analyzed data from BLM's Abandoned Mine Lands/Site Cleanup Inventory database as of April 28 and 29, 2014. In 2005, Interior's Office of Inspector General (OIG) found that this inventory data were not complete and reliable and recommended that BLM validate existing inventory data and develop procedures for ongoing data collection to ensure that data in the inventory are complete, accurate, and consistent. The Interior OIG has not yet closed this recommendation. According to BLM, these data are an authoritative source of information regarding the inventory and status of sites. According to BLM's website, BLM is constantly reviewing and updating its data. In addition, the majority of sites were reported during the initial inventory efforts from various sources, but they have not been assessed on the ground by BLM.

We also collected information from the U.S. Army Corps of Engineer's Formerly Used Defense Sites (FUDS) Management Information Systems on former defense sites located on land managed by USDA and Interior. We are reporting the inventory number that is currently in that system. To determine whether the data were sufficiently reliable for reporting the number of USDA and Interior FUDS sites, we provided the inventory lists to USDA and Interior officials, who confirmed that the data accurately reflected their FUDS inventories. Because the data from the U.S. Army Corps of Engineers (Corps) were not used to support any findings, conclusions, or recommendations in our report, we did not review how the Corps developed its inventory list.

Funding

To describe the funding levels for identifying and addressing contaminated sites, we collected budget and expenditure data for fiscal years 2003 through 2013 from responsible agency officials. Specifically, we collected funding data and project lists for the following:

- USDA's department-wide Hazardous Materials Management Appropriations account;
- Forest Service's Environmental Compliance and Protection program, as well as USDA's one-time Recovery Act funds and cost recovery funds;
- funds the CCC is authorized to borrow;
- Interior's department-wide Central Hazardous Materials Fund;
- BLM's Abandoned Mine Lands subactivity account;
- BLM's Hazardous Materials Management subactivity account;
- NPS's cleanup fund; and
- the Fish and Wildlife Service's Refuge Cleanup Fund.

To better understand how USDA's environmental restoration funds were used, we interviewed officials from each of USDA's responsible agencies. At Interior, we met with officials from Interior's Office of Environmental Policy and Compliance, Environmental Cleanup and Liability Management. We also interviewed officials with BLM, NPS, and Fish and Wildlife Service about their cleanup funds. We reviewed policies for awarding cleanup project funds and lists of project funding. We also collected data, as of September 30, 2013, from Interior's EDL database on the stage of cleanup of each site in the system.

Finally, we interviewed agency officials regarding efforts to develop a comprehensive estimate of the cost to address environmental contamination at abandoned mines on Forest Service, BLM, and NPS lands.

Environmental Liabilities

To describe the amount of environmental liabilities USDA and Interior reported on its financial statements, we reviewed the Chief Financial Officer's Act of 1990 and the Federal Accounting Standards Advisory Board's *Statement of Federal Financial Accounting Standards 5: Accounting for Liabilities of the Federal Government* (SSFAS No. 5) and its supplemental guidance, *Technical Release 2: Environmental Liabilities Guidance*.[1] We reviewed the policies, procedures, and methodologies the departments provide their agencies to estimate their environmental liabilities. When available, we reviewed the policies and procedures for estimating and reporting environmental liabilities from each agency or bureau. In addition, we interviewed officials in both Interior's and USDA's Office of the Chief Financial Officer, Interior's Office of the Solicitor, and USDA's Office of General Counsel. Furthermore, we reviewed the fiscal year 2013 consolidated financial statements of both USDA and Interior. Because USDA's fiscal year 2013 financial statements were audited by the USDA OIG, and Interior's 2013 financial statements were audited by an independent public accounting firm and determined to be reliable, we reviewed the statements but did not perform additional audit procedures to verify the completeness or accuracy of the amounts reported. We met with officials from ARS, CCC, and the Forest Service to understand their process for selecting sites for inclusion as environmental liabilities on their annual financial statements. At Interior, we reviewed information from the EDL database and interviewed officials with Interior's Office of Environmental Policy and Compliance, Environmental Cleanup and Liability Management. This office develops the environmental liability policy for Interior and maintains the EDL database for documenting liabilities. We also reviewed Interior's *Environmental and Disposal Liabilities Identification, Documentation and Reporting Handbook* and agency handbooks for implementing the process.

EPA's Role

To determine EPA's role with regard to USDA and Interior sites, we obtained and reviewed relevant laws, regulations, and agency policies. We interviewed officials with EPA's Federal Facilities Restoration and Reuse Office and USDA and Interior headquarters officials on the procedures for listing federal sites on the docket and efforts to take steps to assure that agencies conduct preliminary assessments and any other cleanup-related activities. We selected three EPA regions with the most USDA and Interior sites on the docket to interview about the process for listing federal sites. However, we did not interview EPA docket officials in California because the docket coordinator had retired. We documented the cleanup status of the four USDA and Interior NPL sites.

EPA published the original docket in 1988 and has published additions, deletions, and corrections 26 times. EPA provided a copy of the spreadsheet it has compiled that lists the USDA and Interior docket sites. EPA recently compared the spreadsheet with each docket publication. In addition, it provided the compiled docket to each EPA regional office for our review. Because of this recent verification, we believe the docket is sufficiently reliable for reporting the number of docket sites EPA has published. We also obtained each updated docket EPA has published in the *Federal Register* to determine how often EPA has made such updates.

To determine whether the USDA and Interior agencies have conducted preliminary assessments of sites on EPA's docket, we collected and analyzed data from the Comprehensive Environmental Response, Compensation, and Liability Information System as of February 2014. Because of the age of the data and its potential unreliability, we asked Interior and USDA to review the data in EPA's database relating to sites that EPA had not noted had met the preliminary assessment requirement. Both USDA and Interior noted the sites that they believed had met the preliminary assessment requirement. Because we are reporting the information currently in the system—and also providing data on the discrepancies USDA and Interior report with regard to this data—we believe the information is sufficiently reliable for reporting.

Finally, for all four objectives, we reviewed relevant laws—in particular, CERCLA and RCRA—and associated regulations and directives.

We conducted this performance audit from April 2013 to January 2015 in accordance with generally accepted government auditing standards. Those standards require that we plan and perform the audit to obtain sufficient, appropriate evidence to provide a reasonable basis for our findings and conclusions based on our audit objectives. We believe that the evidence obtained provides a reasonable basis for our findings and conclusions based on our audit objectives.

APPENDIX II: NUMBER OF AND CLEANUP COSTS FOR FORMERLY USED DEFENSE SITES ON DEPARTMENT OF THE INTERIOR AND U.S. DEPARTMENT OF AGRICULTURE PROPERTIES

The Department of the Interior (Interior) and the U.S. Department of Agriculture (USDA) both manage lands that contain formerly used defense sites (FUDS). Specifically, FUDS are defined as properties that were under the jurisdiction of the Department of Defense (DOD) and owned by, leased to, or otherwise possessed by the United States at the time of actions leading to contamination by hazardous substances or other hazards prior to October 17, 1986. Under the Defense Environmental Restoration Program, DOD is required to carry out a program of environmental restoration activities at FUDS.[1] Authority for cleaning up FUDS was delegated to the Department of the Army, which in turn delegated it to the U.S. Army Corps of Engineers (Corps). According to Army officials, the Corps may, as appropriate, address FUDS located on lands managed by Interior and USDA in consultation with these two departments.[2] As of the end of fiscal year 2012, the Corps had identified 735 and 236 eligible and approved FUDS projects on Interior and USDA properties, respectively. About 46 percent of these projects were designated as "no DOD action indicated" (NDAI). According to Corps officials, NDAI signifies a FUDS property or project that meets any of these conditions: (1) is not eligible for FUDS program consideration; (2) is categorically excluded from the FUDS program; (3) contains hazards that were not the result of DOD actions on or before 17 October 1986, pose no threat to human health or safety or the environment, and require no additional environmental restoration activities; and (4) for which the required response action has been completed.[3]

FUDS projects generally fall under one of three program categories: Installation Restoration Program, Military Munitions Response Program, and Building Demolition/Debris Removal Program.[4]

- **Installation Restoration Program.** This program category includes FUDS with (1) hazardous, toxic, and radioactive waste and (2) containerized hazardous, toxic, and radioactive waste, such as aboveground and underground storage tanks. Projects under this program category focus on addressing releases of hazardous substances or pollutants and contaminants; petroleum, oil, or lubricants; DOD-unique materials; hazardous wastes or hazardous waste constituents; low-level radioactive materials or low-level radioactive wastes; and explosive compounds released to soil, surface water, sediments, or groundwater as a result of ammunition or explosives production or manufacturing at ammunition plants. The Corps categorized 317 FUDS projects on Interior properties and 102 FUDS projects on USDA properties under this program, as of the end of fiscal year 2012. [5] Of these projects, 190 and 67, respectively, were designated NDAI.

- **Military Munitions Response Program.** This program category includes FUDS with munitions and explosives of concern, munitions constituents, and recovered chemical warfare materiel. 6 The Corps categorized 367 FUDS projects on Interior properties and 106 FUDS projects on USDA properties under this program, as of the end of fiscal year 2012.[7] Of these projects, 91 and 36, respectively, were designated NDAI.
- **Building Demolition and Debris Removal Program.** Projects under this program category focus on the demolition and removal of unsafe buildings and structures and the removal of unsafe debris from eligible FUDS properties. To be eligible, the property's conditions must have been hazardous as a result of prior DOD use and must have been inherently hazardous (i.e., present a clear danger, likely to cause, or have already caused, death or serious injury to a person exercising ordinary and reasonable care) when the property was transferred by DOD. The Corps categorized 47 FUDS projects on Interior properties and 23 FUDS projects on USDA properties under this program, as of fiscal year 2012. Of these projects, 39 and 20, respectively, were designated NDAI.

The Corps reports cleanup costs of over $405 million at FUDS on Interior properties and $223 million at FUDS on USDA properties over the past decade. These amounts include costs expended during fiscal years 2003 through 2012 and costs allocated for fiscal year 2013.[8] The Corps estimates it will cost an additional $3.3 billion to complete FUDS projects on Interior properties and $1.4 billion on USDA properties. Both past and future costs are for cleanup activities at FUDS that have undergone a preliminary assessment and where further action is warranted.[9] According to Corps officials, a FUDS site or project area may have several landowners or property managers; therefore, FUDS develops and reports costs on a site specific or project specific basis, regardless of landowner. This means that the costs reported above may overestimate the actual cleanup costs on DOI or USDA properties as an individual project may only be partially located on federally managed land.

APPENDIX III: CLEANUP STATUS OF DEPARTMENT OF THE INTERIOR'S ENVIRONMENTAL AND DISPOSAL LIABILITY SITES

Department of the Interior's (Interior) bureaus are directed to record[1] their environmental and disposal liabilities (EDL) in the EDL database module of the department's Environmental Management Information System on a quarterly basis.[2] In order for Interior and its bureaus to track the progress of cleanup at EDL sites, the database prompts the bureaus to record the current stage of the cleanup process for each EDL site. One of five stages, as described by Interior below, can be recorded.

- **Due care only.** An EDL site is recorded as having "due care only" when no activity beyond an environmental assessment has been completed at the site.[3]
- **Study.** This stage of the cleanup process is associated with an EDL site that has had due care completed and where the level of contamination is significant enough to warrant further study or future cleanup. During this stage, bureaus may complete a preliminary assessment/site investigation, remedial investigation/feasibility study, or other investigation study at the EDL site. Such studies are used, for example, to determine the nature and extent of contamination, assess the treatability of site contamination, and evaluate the potential performance and cost of treatment technologies.
- **Cleanup/remediation/removal**. This stage of the cleanup process is associated with a cleanup response action (removal or remedial) at an EDL site.
- **Operation and maintenance**. This stage of cleanup process is associated with an EDL site requiring operation and maintenance of, for example, cleanup systems dedicated to a cleanup response action. Operation and maintenance is applicable after the remedial action has achieved cleanup action goals and is determined to be operational and

functional, or 1 year after construction (whichever is earlier) except for water treatment alternatives.[4]

- **Long-term monitoring.** This stage of the cleanup process is associated with an EDL site that requires monitoring to minimize and control the contamination. For example, the monitoring of groundwater may be required to ensure capture or natural attenuation is occurring.

As of the end of fiscal year 2013, about 40 percent of the bureaus' 756 EDL sites were recorded as being in the study stage of the cleanup process. The second and third highest recorded stages, respectively, were cleanup/remediation/removal (22.9 percent), followed by due care only (18.9 percent). The number of EDL sites in each stage of the cleanup process is shown below in figure 4.

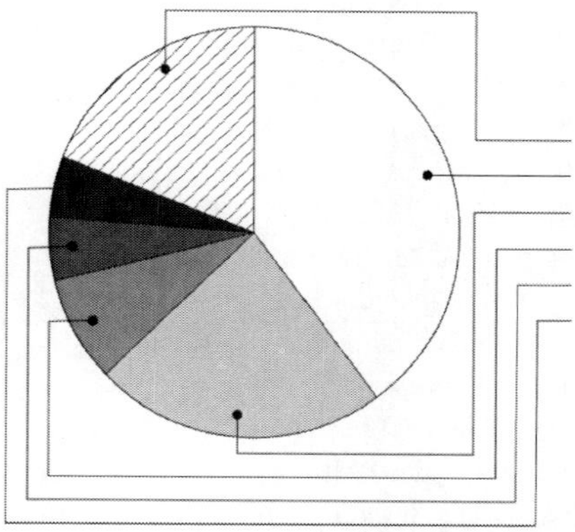

Stage	Percentage	Number of EDL sites
Due care only	18.9%	143
Study	39.9%	302
Cleanup/remediation/removal	22.9%	173
Operation and maintenance	8.3%	63
Long-term monitoring	5.0%	38
Not recorded[a]	4.9%	37

EDL = Environmental and Disposal Liability.

Source: GAO analysis of Department of the Interior data and its responses to follow-up data questions. | GAO-15-35.

[a]According to bureau officials, of the 37 EDL sites with no cleanup stage recorded, 1 site required no further action; 1 site is in multiple stages of the cleanup process; and 35 sites were input into the database before the standards for due care were established but the responsible bureau is currently in the process of hiring an engineering contractor to conduct due care to either establish or refute the contamination liability at the sites.

Figure 4. Number of Environmental and Disposal Liabilities (EDL) Sites in Each Stage of the Cleanup Process, as of Fiscal Year 2013.

APPENDIX IV: USDA AND INTERIOR NATIONAL PRIORITIES LIST SITES

This appendix provides the cleanup status of the two U.S. Department of Agriculture (USDA) sites and two Department of the Interior (Interior) sites currently on the National Priorities List (NPL). (See table 5.)

Table 5. USDA and Interior NPL Sites, as of May 2014

USDA NPL sites	
Agricultural Research Service (ARS) Beltsville Agricultural Research Center	This site is a large agricultural research complex spanning 6,500 acres in Beltsville, Maryland. The site was listed on the NPL due to contamination as a result of past disposal practices. Contaminants include polychlorinated biphenyl, pesticides, metals,radioactive materials, and solvents in soil and groundwater. In 1998, USDA and Environmental Protection Agency (EPA) finalized a Federal Facility Agreement under which they agreed to, among other things, establish a process for investigating areas of concern, identify response actions for these contaminated areas, and clean them up. USDA originally identified 167 areas of concern. As of February 2014, 13 such areas remain where assessment or cleanup is anticipated.
Forest Service White King/Lucky Lass, Oregon	This site is the location of two abandoned uranium mines located approximately 1 mile apart. The site encompasses approximately 140 acres affected by uranium mining activities that occurred during the 1950s and 1960s. The site is contaminated with arsenic and radionuclides in water-filled excavation pits (ponds) and stockpiled mineralized waste piles. EPA listed the site on the NPL in 1995. That same year EPA and Forest Service entered into an Administrative Order on Consent with several private sector potentially responsible parties (PRP). PRPs subsequently performed investigations and remedial actions at the site. EPA and the Forest Service also signed a Memorandum of Understanding in 1995. A group of private PRPs constructed the remedy under a 2005 Consent Decree. Construction of all remedies at the site is complete. In 2010, EPA conducted a Five Year Review at the site and noted that the remedial actions at the site are complete and protective of human health and the environment.

Interior NPL sites	
Bureau of Land Management (BLM) Lee Acres Landfill, New Mexico	In 1962, BLM leased a portion of land to San Juan County to operate a county landfill. The landfill disposed of liquid waste in lagoons. In 1985, a lagoon was breached. The county closed the landfill and covered a lagoon on site with soil. Volatile organic compounds were found in three domestic water supply wells in a nearby subdivision. BLM distributed bottled water to some local residents and ultimately connected them to the town water supply. EPA listed the site on the NPL in 1990. Contaminants of concern include nickel, vinyl chloride, trichloroethene, and manganese. In 1993, BLM,EPA and the New Mexico Environment Department entered into a technical Memorandum of Understanding for completion of a remedial investigation. After further study and approved plans, EPA and Interior signed an interagency agreement in 2004. The remedy construction was completed at the site in 2005 with the construction of a landfill cover and monitoring wells. The groundwater is to be restored through natural attenuation. In 2009, BLM prepared a Five Year Review of the site. The remedial actions performed at the site are considered to be protective of human health and the environment. BLM has plans to conduct a second 5 year review in 2014.
Fish and Wildlife Service Crab Orchard National Wildlife Refuge, Illinois	The site was a location of a bomb and explosive plant during World War II and was established by the federal government as a refuge in 1947. Industries continued to lease the site, and some used unlined landfills and dumps to dispose of waste from their operations. The site has polychlorinated biphenyl, cadmium, and lead contamination. EPA listed it on the NPL in 1987. In 1991, EPA, Illinois EPA, Interior, and the Department of the Army signed a Federal Facility Agreement. A number of cleanup remedies have been implemented at the site, including the excavation and treatment of contaminated soils, the installation of a pump and treat system for one groundwater plume, and monitoring of a second plume for natural attenuation. In June 2011, EPA signed the fourth Five Year Review report for the site and concludedthat the site is expected to protect human health and the environment.

Sources: GAO analysis of documents from USDA and Interior. \| GAO-15-35.

Appendix V: Comments from the U.S. Department of Agriculture

USDA

United States Department of Agriculture

Office of the Assistant Secretary for Administration

Office of Procurement and Property Management

300 7th Street Southwest Room 302 Reporters Building

Washington, DC 20024-9300

Mr. Vincent Price
Assistant Director,
Natural Resources and Environment
Government Accountability Office
Washington, DC 20548

Dear Mr. Price:

The U.S. Department of Agriculture (USDA) appreciates the opportunity to respond to the U.S. Government Accountability Office (GAO) draft report concerning hazardous waste sites on USDA and Interior Lands, dated January 2015.

USDA does not agree with the key findings in the GAO draft report. We are providing (1) a narrative that clarifies USDA's position on the key findings (Attachment 1) and (2) a comment matrix with suggested word changes for the final report (Attachment 2).

Thank you again for the opportunity to review and respond to the GAO draft report.

Sincerely,

Lisa M. Wilusz
Director
Office of Procurement and Property Management

Attachment 1

USDA Response to GAO Key Findings from Draft Report Concerning Hazardous Waste Sites on USDA and Interior Lands

This document describes the status of the USDA centralized hazardous waste site inventory and presents USDA views on GAO findings related EPA and USDA responsibilities associated with the identification of the cleanup of contaminated sites.

GAO's Inventory-Related Findings

(1) The GAO report states that "USDA does not have a centralized site inventory or plans and procedures for completing one, in particular, for abandoned mines on some USDA lands" ("Introductory/Preface" page). Page 7 of the report similarly states that USDA "has not developed a complete, consistent, or usable inventory of abandoned mines and has no plans and procedures for developing an inventory".

USDA Response:

As USDA staff explained during interviews with GAO staff related to this report, the USDA centralized inventory is in a transition phase as a result of reduced funding levels. USDA experienced significant yearly budget cuts from $15.7 million to $3.7 million, amounting to a 76 percent reduction over the past ten years. These budget cuts significantly affect the way we manage our program and forced us to answer some very difficult questions. For example:

- Can we afford a centralized inventory?
- What benefits will a centralized inventory provide?
- Based on historical funding, how can we afford to identify every potentially contaminated mine site on Forest Service Managed lands?

In answering these questions, USDA decided to maintain the database, which contains our centralized inventory, but modify it in an effort to reduce operating costs.[1] We also looked at the most cost effective way to house the data. As a result, we moved this database to a cloud environment, instead of dedicated servers. Finally, we looked at ways to reduce redundancy in data collection across USDA agencies. When looking at the data, we found that abandoned mine lands (AML) make up the bulk of the potentially contaminated sites inventory. We also recognized that AML pose a multitude of diverse challenges such as protection of bat habitats and physical safety hazards. Moreover, only a subset of this entire inventory will result in contamination levels which will require a cleanup action.

With these issues in mind, as well as Forest Service efforts to develop their own agency-wide database to address some of these other challenges associated with AML, USDA decided to move the AML portion of the centralized inventory to the Forest Service. This decision represents a significant cost savings for USDA and reduces redundant data collection. USDA will continue to track any sites within the USDA database where cleanup is deemed

---

[1] The USDA centralized inventory referenced in the GAO report contains 1,491 contaminated sites (page 6, line 3) and 20,401 potentially contaminated sites (page 8, 2nd paragraph, line 4).

necessary. Despite some difficulties with this transition plan, USDA is moving forward with it. USDA agrees that our centralized inventory requires some updating due to this transition; however a centralized inventory in need of some updating is nonetheless an existing centralized inventory.

(2) The GAO conclusions further state (p. 33): "Without a comprehensive, reliable, and easily accessible inventory of potentially contaminated sites on lands managed by USDA, including abandoned mines, and plans and procedures for developing such an inventory, USDA cannot either effectively manage the department's cleanup programs or ensure that limited funds are targeted to cleanup those site that may pose the greatest threats to human health or the environment".

USDA's Response:

USDA is effectively managing its cleanup program within the given constraints of limited resources. Although, we recognize that our potentially contaminated inventory is not complete, we strongly disagree that an incomplete inventory leads to ineffective management of the program, as stated in the GAO report.

USDA recognizes that resources are not, nor are they likely to become, available to complete an on-the-ground inventory of all USDA-managed lands in the foreseeable future. Therefore, the USDA AML inventory strategy includes utilizing data from all existing sources and development of collaborative partnerships with Federal and State agencies. Inventory and cleanup efforts are focused on State and/or local priority areas, such as high-visitation areas and threatened watersheds. Cleanup efforts are initiated at sites with the most acute public safety and/or environmental hazards.

A more focused USDA AML inventory and cleanup approach also supports the Department's strategic plan with respect to protecting and enhancing America's water resources. As water continues to become a more valuable resource, we believe that an effort made towards cleaning up priority watersheds is the best use of limited resources. A holistic watershed approach unifies all stakeholders (including private interests groups and Native American Tribes) in the identification, prioritization, and cleanup of important drinking water supplies, fisheries, and wildlife habitat within an entire watershed. Such partnerships have been instrumental in the progress achieved to date.

Focusing our future AML inventory efforts in areas determined to be high priorities through coordination and collaboration reduces costs through leveraging funds. Such efforts also achieve effectiveness in facilitating solutions to address mixed ownership issues, avoid duplication of efforts and conflicting actions, and assist in identifying those sites that pose the highest potential for harm to public health and safety.

GAO Findings with Respect to EPA and USDA Responsibilities

(1) The GAO report states that "Disagreements with USDA and Interior over their need to assess the remaining sites and differing information on whether this requirement has been met at some sites mean EPA cannot assure that the assessments are conducted in a timely manner and, ultimately, that sites most in need of remediation are addressed" (preface & page 33).

USDA's Response:

USDA has delegated CERCLA authority with respect to land under its jurisdiction, custody, or control except that EPA retains authority in cases of emergencies and for remedial actions at sites that are listed on the National Priorities List (NPL). Therefore USDA, not EPA, sets the priority for response actions at non-NPL sites on USDA-administered land. Specifically the disagreements referenced in the report occur in cases where (1) there is no evidence that a release has occurred, (2) the site location is not defined by EPA to the extent feasible to do an assessment (i.e., docketing an entire national Forest), and (3) EPA did not provide USDA with enough information to identify the location of the site. USDA acknowledges the need to improve coordination with EPA with respect to these situations, and has already begun face-to-face meetings with each EPA Region to address these issues. Disagreements between EPA and USDA, however, do not ever delay needed remediation work.

(2) The report also states that "USDA and Interior generally do not perform preliminary assessment if they do not agree with EPA's determination that a facility should have been added to the docket" (Page 29).

USDA's Response:

With regard to the failure to perform preliminary assessments at disputed sites on the EPA docket, it is important to understand how sites get listed on EPA's docket. The docket contains information submitted to EPA by federal agencies under the following authorities: CERCLA Section 103 and the Resource Conservation and Recovery Act (RCRA) Sections 3005, 3010, and 3016. GAO's comments about USDA not performing preliminary assessments if they do not agree with EPA's determination generally relate to sites listed on the docket pursuant to RCRA. While we acknowledge EPA's authority to place these sites on the docket, USDA does not believe that facilities in compliance with RCRA, which must generate waste disposal manifests, should be required to conduct CERCLA investigations as a result of docket listing, particularly where there is no record of a release of hazardous substances into the environment. The result of this practice would be unnecessary expense with essentially no benefit. As a matter of policy, EPA should develop an alternative, less wasteful process. In almost all cases, these sites have not had a release of hazardous substances and, therefore, do not pose a threat to the environment, human health, or safety.

(3) In addition, the report explains how sites are scored by EPA and categorized (pages 30-31) without adequately defining what a no further action determination by EPA means to other federal agencies.

USDA's Response:

For docket purposes, a no further remedial action planned (NFRAP) letter from EPA means that EPA has determined the site does not require NPL listing. The vast majority of USDA sites do not warrant NPL listing and a docket NFRAP letter does not preclude USDA from taking action. The threat may be significant and cleanup by USDA warranted. In those cases, USDA prioritizes its sites and proceeds with cleanup using its delegated CERCLA authorities with little or no EPA involvement.

APPENDIX VI: COMMENTS FROM THE ENVIRONMENTAL PROTECTION AGENCY

UNITED STATES ENVIRONMENTAL PROTECTION AGENCY
WASHINGTON, D.C. 20460

DEC 9 2014

OFFICE OF
SOLID WASTE AND
EMERGENCY RESPONSE

Mr. Alfredo Gomez
Acting Director
Natural Resources and Environment
U.S. Government Accountability Office
Washington, DC 20548

Dear Mr. Gomez:

Thank you for the opportunity to review and comment on GAO's draft report, "Hazardous Waste: Agencies Should Take Steps to Improve Information on USDA and Interior's Potentially Contaminated Sites" (GAO-15-35).

The purpose of this letter is to provide the Environmental Protection Agency's response to your recommendations. The EPA generally agrees with the GAO's findings, conclusions, and recommendations.

In its review, GAO found that the Departments of Agriculture and the Interior do not have complete inventories of contaminated or potentially contaminated sites on their properties. Both Departments' estimates of their financial liabilities for land cleanup were found to be reasonable. GAO found that the EPA's primary role is to maintain a list of potentially contaminated sites on USDA and Interior land, based on data provided by the agencies; to ensure the agencies assess the sites; and to oversee cleanup as needed. The report states that the EPA's ability to ensure that preliminary assessments are conducted in a timely manner and that sites in most need of remediation are addressed is inhibited by disagreement from the two agencies about what cleanup work is necessary and whether the preliminary assessment requirement has been met. The report also noted that the EPA has an ongoing initiative to work with states and federal agencies on completing outstanding assessments.

The EPA believes that the recommendations contained within this report will help to address the 50 USDA sites and 79 Interior sites whose status is being disputed; moreover, more may be needed for other contaminated or potentially contaminated sites on USDA and Interior lands, as the EPA cannot compel agencies to comply with requests for site assessments to be completed.

GAO Recommendations

- Review available information on USDA and Interior sites where the EPA's Superfund Enterprise Management System indicates that a preliminary assessment has not occurred to determine the accuracy of this information, and update the information, as needed;

- After completing this review, inform USDA and Interior whether the requirement to conduct a preliminary assessment at the identified sites has been met or if additional work is needed to meet this requirement; and
- Work with the relevant USDA and Interior offices to obtain any additional information needed to assist the EPA in determining the accuracy of the agency's data on the status of preliminary assessments for these sites.

EPA Response

The EPA agrees with the recommendation to review the status of USDA and Interior sites to identify if a preliminary assessment is still needed. The EPA will compare the site data provided by USDA and Interior during the review to existing Superfund Enterprise Management System data in order to identify these sites.

The EPA agrees with the recommendation to inform USDA and Interior whether the requirement to conduct a preliminary assessment has been met or if additional work is needed. The EPA will use the data collected in response to the first executive recommendation to inform the other Federal agencies of their statutory requirements to complete site assessment work.

The EPA agrees with the recommendation to work with USDA and Interior offices to obtain additional information to determine the accuracy of the agency's data on the status of preliminary assessments for these sites.

Comments about terminology and phrasing in the report have been included separately in the document itself (enclosed).

During the EPA's Federal Facility Cleanup Dialogue held with the Department of the Interior and Department of Agriculture on, October 21, 2010, the EPA encouraged USDA and DOI to work with the EPA on the need to promote transparency, greater collaboration and joint problem solving; have a conversation about challenges of federal facility cleanups; clarify federal roles and responsibilities; and develop a longer-term strategy for addressing key issues. The EPA also recommended that DOI and USDA should consider the Department of Defense, Department of Energy, and other agencies to determine what lessons learned and best practices can be adapted to their issues. The EPA also recognized the need for DOI and USDA to establish baseline facts and develop an inventory of known sites in order to advance site cleanup.

On behalf of the EPA, I would like to thank you for the opportunity to review the draft report. If you require further information, please contact Ellen Treimel at (703) 603-0720.

Sincerely,

Mathy Stanislaus
Assistant Administrator

Enclosure

cc: EPA GAO Liaison Team
Barry Breen
Nitin Natarajan

Becky Brooks
Charlotte Bertrand
Paul Leonard
Charles Openchowski
Lora Culver
Bobbie Trent
Johnsie Webster
Gwendolyn Spriggs
Kecia Thornton
Nick Hilosky
Holly Fliniau
Mark Howard
Willie Taylor
Lisa Wilusz

End Notes

[1] In this report, we use the term "hazardous waste" to refer generally to material that is or may be harmful to human health or the environment, although the term has specific meaning under the Resource Conservation and Recovery Act (RCRA) and the Comprehensive Environmental Response, Compensation, and Liability Act (CERCLA). Under RCRA, "hazardous waste" generally refers to materials specifically listed by the Environmental Protection Agency or which demonstrate certain hazardous characteristics. A "hazardous waste" under RCRA is also among the substances defined as a "hazardous substance" under CERCLA.

[2] Pub. L. No. 101-576, 104 Stat. 2838 (codified as amended in scattered sections of 5, 31, 28, and 42 U.S.C.).

[3] Federal Accounting Standards Advisory Board's Statement of Federal Financial Accounting Standards (SFFAS) No. 5, *Accounting for Liabilities of the Federal Government* (Dec. 20, 1995).

[4] GAO, *Superfund: Civilian Federal Agencies Slow to Clean Up Hazardous Waste,* RCED87-153 (Washington, D.C.: July 24, 1987) and *Status of Civilian Federal Agency Efforts To Address Hazardous Waste Problems on Their Lands*, RCED-84-188 (Washington, D.C.: Sept. 28, 1984).

[5] GAO, *Hardrock Mining: Information on Abandoned Mines and Value and Coverage of Financial Assurances on BLM Land*, GAO-08-574T (Washington, D.C.: Mar. 12, 2008); *Abandoned Mines: Information on the Number of Hardrock Mines, Cost of Cleanup, and Value of Financial Assurances*, GAO-11-834T (Washington, D.C.: July 14, 2011); and *Hardrock Mining: BLM Needs to Revise Its Systems for Assessing the Adequacy of Financial Assurances*, GAO-12-189R (Washington, D.C.: Dec. 12, 2011).

[6] Reporting is also referred to as recognizing a liability.

[7] Under CERCLA, potentially responsible parties include current or former owners or operators of a site or the generators and transporters of the hazardous substances. For purposes of this report, we use the term responsible parties to refer to those potentially responsible parties who are accepting liability or for whom liability is proven.

[8] Federal Accounting Standards Advisory Board, Statement of Federal Financial Accounting Standards (SSFAS), No. 5, *Accounting for Liabilities of the Federal Government* (Dec. 20, 1995).

[9] Federal Accounting Standards Advisory Board, Federal Financial Accounting and Auditing Technical Release 2, *Determining Probable and Reasonably Estimable for Environmental Liabilities in the Federal Government.*

[10] Disclosure refers to information in the notes regarded as integral to the financial statements.

[11] In addition, certain Formerly Used Defense Sites (FUDS) are now under USDA or Interior's jurisdiction. These sites were once under the jurisdiction of the Department of Defense and may be contaminated. They include firing ranges, and industrial facilities. Appendix II provides information on FUDS.

[12] The National Forest System (NFS) is a national system of federally owned units of forest, range, and related land that are administered by the Forest Service or designated for administration through the Forest Service.

[13] These sources include the Environmental Management Division's (EMD) Center for Environmental Excellence (CEE) database for Forest Service sites in Regions 1 through 9, a spreadsheet of Forest Service sites for Region 10, USDA's Docket Status Spreadsheet, and statements from officials of the Animal and Plant Health Inspection Service (APHIS), Agricultural Research Service (ARS), and Commodity Credit Corporation (CCC).

[14] However, Forest Service officials said that some landfills and underground storage tanks may not be captured by this number. The Forest Service tracks most of its sites with confirmed contamination in the CEE database. In general, sites brought into "active" status within CEE have known contamination, but the degree and extent of that contamination has yet to be determined. Active sites are those at which USDA plans to perform (or has performed) a CERCLA preliminary assessment to eliminate from further consideration those sites that pose no threat to public health or the environment, to determine the need for further response actions, and to set priorities for further response actions.

[15] Forest Service officials said that these types of sites are sometimes addressed through time-critical removal actions and are cleaned up shortly after their discovery, depending on the specific circumstances.

[16] According to USDA officials, this estimate is derived from a 1995 USDA report that acknowledged uncertainties associated with the date used to generate the estimate. For some states, the data are fairly accurate but, in other cases, the data are incomplete or missing.

[17] In CEE, sites that are marked as "pending" have not yet been assessed and are therefore potentially contaminated. CEE shows that there are 20,401 "pending" mine or mill sites.

[18] USDA officials said that CEE still contains the original abandoned mine lands inventory and will do so until the AML database is completed. However, additional inventory work has been performed during the transition to the AML database, and the completeness of the data varies across the regions. Therefore, the abandoned mine data within CEE represent a subset of the data in the AML database.

[19] According to Forest Service officials, the Forest Service regional offices conducted inventory efforts in the 1980s and 1990s. Officials noted, however, that each regional office inventoried its sites differently, and some inventory efforts were more comprehensive than others.

[20] Officials in Forest Service Region 10 said that their region's focus has been on cleanup of known priority sites rather than completing their inventory of potentially contaminated sites.

[21] Forest Service Region 2 has helped states in its region, most notably Colorado, develop their own inventories of abandoned mines, but Forest Service Region 2 does not maintain inventory data at the regional level and was only able to provide inventory data from CEE—which contains a more limited data set. To provide a full inventory of sites in the region, one Forest Service official told us that it would have to get the data, much of it in

hard copy, from each ranger district—a very time-consuming task. Officials in Forest Service Region 2 added that their management approach is to spend their limited resources working with other state and federal agencies to address agreed-upon areas of concern rather than completing their inventory efforts.

[22] A mine feature is a single human-made object or disturbance associated with mining. A mine site can be composed of one or more features.

[23] The remainder of the hazardous materials sites has been remediated or requires no further action.

[24] U.S. Department of Interior Office of Inspector General, Audit Report: Abandoned Mine Lands in the Department of the Interior, July 2008, C-IN-MOA-0004-2007.

[25] In May 2013, BLM published a report that summarizes recent inventory efforts titled *Abandoned Mine Lands: A New Legacy.*

[26] BLM has 12 state offices located in Alaska, Arizona, California, Colorado, Idaho, Montana, Nevada, New Mexico, Oregon, Utah, Wyoming, and an eastern states office.

[27] NPS published interim results in Interim Inventory and Assessment of Abandoned Mineral Lands in the National Park System, Natural Resource Technical Report NPS/NRSS/GRD/NRTR-2013/659, January 2013. Complete results, as shown above, will be published in a comprehensive report to be released later in 2014.

[28] American Recovery and Reinvestment Act of 2009, Pub. L. No. 111-5, 123 Stat. 115.

[29] This total includes overhead costs; $11.85 million was for site projects in fiscal year 2013.

[30] Several USDA agencies may also use agency specific funds for environmental cleanup activities, but those expenditures are usually a small percentage of the total budget and are therefore excluded from our analysis. For example, APHIS's Environmental Protection Program may fund some environmental cleanup projects. However, one APHIS official told us that APHIS does not receive any specific appropriation for environmental cleanup efforts, and that most of the funds in this program actually go toward disposal of hazardous waste from operating facilities. FSA also funds its cleanup of foreclosure properties through the Farm Loan Program. According to one FSA official, in fiscal year 2013 FSA spent approximately $1.47 million on cleanup of its three foreclosure properties out of the Farm Loan Program's $4.8 billion budget.

[31] EMD provides funds to the Office of the General Counsel (OGC) to support USDA cleanup efforts. EMD officials said that OGC has been successful in bringing additional money into the HMMA program by, among other things, identifying potentially responsible parties (PRP), such as owners and operators of a site, that may help pay for cleanup activities and by pursuing cost recovery actions against these PRPs. These officials added that—unlike Interior, which pays for staff salaries out of general appropriations—USDA pays for its staff salaries out of program specific accounts.

[32] These funds include overhead costs; for example, in fiscal year 2013, $11.85 million was for cleanup projects.

[33] 15 U.S.C. § 713a-4 (2012).

[34] USDA's cost recovery amounts include (1) costs already incurred by USDA and reimbursed by a PRP, (2) funds that a PRP has agreed to provide for future cleanup work, and (3) the value of cleanup work directly performed by the PRP.

[35] USDA officials noted that this figure only includes assessment activities conducted after 2003. USDA officials told us that the agency assessed and cleaned up many sites prior to the establishment of CEE that have not been recorded in that database.

[36] According to USDA officials, this number represents the number of CERCLA cleanup actions in CEE, as well as certain RCRA or UST sites. Several sites had more than one cleanup activity.

[37] EMD also issued accomplishment reports prior to 2011. However, according to agency officials, in 2011, EMD changed the format of the report to include background information to reach a broader audience and serve as a program education tool for individuals not familiar with the cleanup program.

[38] BLM has 12 state offices located in Alaska, Arizona, California, Colorado, Idaho, Montana, Nevada, New Mexico, Oregon, Utah, Wyoming, and an eastern states office.

[39] USDA did not report asbestos cleanup liabilities prior to fiscal year 2013. As a result, USDA only reported $10 million in environmental and disposal liabilities in fiscal year 2012.

[40] CCC reported a cost range of $8 million to $55 million on its agency financial statements.

[41] These 24 Forest Service sites do not include any abandoned mines.

[42] A contingent liability is an existing condition, situation, or set of circumstances involving uncertainty as to possible loss to an entity that will ultimately be resolved when one or more future events occur or fail to occur.

[43] USDA reported $120 million in contingent liabilities in its fiscal year 2013 financial statements; the $40 million in environmental liabilities is a portion of the total reported. The $40 million represents the lower end of a range of cleanup costs for these four mine sites. The mining company that leases these sites sued the U.S. government to recover its response costs incurred in connection with the remediation of the four mines, which are located on federal land. The court ultimately approved a settlement under which the mining company would pay 67 percent, and the government would pay 33 percent of all past and future costs. In court documents, the United States noted that the total cleanup costs for the sites are unknown, but preliminary estimates are in the hundreds of millions of dollars. Nu-West Mining, Inc. v. United States, 2013 U.S. Dist. LEXIS 32747, D. Idaho (2013). Forest Service officials noted that, absent the completion of feasibility studies and the development of remediation cost estimates on these sites, any cost estimates are highly speculative.

[44] USDA did not disclose any similar costs in its fiscal year 2013 financial statements.

[45] According to EPA, the agency also does not include previously owned federal facilities or sole transporters of hazardous waste.

[46] Exec. Order No. 12580, Superfund Implementation, 52 Fed. Reg. 2,923 (Jan. 29, 1987).

[47] We note that the only trigger for the preliminary assessment requirement is inclusion on the docket, not whether a release occurred or whether that release is thought to be low-risk.

[48] 42 U.S.C. § 9620(d)(1) (2012).

[49] These sites may still need cleanup. In those cases, the federal agencies proceed with cleanup using delegated CERCLA authority.

[50] Executive Order 12866 applies to "significant regulatory actions—which include, among other things, actions expected to (1) have an annual effect of $100 million or more on the economy or (2) raise novel legal or policy issues.

[51] Specifically, BLM noted that prior to BLM's implementation of the Federal Land Policy and Management Act of 1976, private operators of mine sites like Red Devil had broad rights to enter and locate mining claims on public lands and to control surface use and activities without federal oversight.

End Notes for Appendix I

[1] Federal Accounting Standards Advisory Board's Statement of Federal Financial Accounting Standards (SSFAS) No. 5, *Accounting for Liabilities of the Federal Government; Federal Financial Accounting Standards 5: Accounting for Liabilities of the Federal Government, Technical Release 2: Environmental Liabilities Guidance.*

End Notes for Appendix II

[1] The Defense Environmental Restoration Program was established by section 211 of the Superfund Amendments and Reauthorization Act of 1986 which amended CERCLA. In implementing this program, DOD is required to carry out its activities addressing hazardous substances, pollutants, or contaminants in a manner consistent with section 120 of CERCLA.

[2] The Army, Interior, and USDA are signatories to the *Statement of Principles for Collaborative Decision Making for Cleanup of Formerly Used Defense Sites on Federal Lands*, in which the agencies agreed to coordinate on FUDS preliminary assessments, and Interior and USDA agreed to consult with DOD on cleanup activities.

[3] Specifically, the Corps designated 320 of the 735 FUDS projects on Interior properties as NDAI, and 123 of the 236 FUDS projects on USDA properties as NDAI.

[4] In addition, the Corps categorized 4 FUDS on Interior properties and 5 FUDS on USDA properties as "community relations" projects. Such projects are associated with public involvement activities and do not fall under one of the three FUDS program categories.

[5] These numbers include "potentially responsible party/hazardous, toxic, and radioactive waste" projects. The Corps defines such projects as FUDS where HTRW cleanup requirements exist, and parties other than DOD are potentially responsible parties for the hazardous substances, pollutants, or contaminants.

[6] Munitions and explosives of concern include unexploded ordnance; discarded military munitions, as defined in 10 U.S.C. § 2710(e)(2); or munitions constituents (e.g., TNT, RDX) present in high enough concentrations to pose an explosive hazard. Munitions constituents are defined by 10 U.S.C. § 2710(e)(4) as materials originating from unexploded ordnance, discarded military munitions, or other military munitions, including explosive and nonexplosive materials, and emission, degradation, or breakdown elements of such ordnance or munitions.

[7] These numbers include "potentially responsible party/military munitions response" projects. The Corps defines such projects as FUDS where Military Munitions Response Program cleanup requirements exist, and parties other than DOD are potentially responsible parties for disposal of the Military Munitions Response Program materials.

[8] A U.S. Army official said the expenditures data the Corps provided to GAO was reported in the last published Defense Environmental Programs Annual Report to Congress, which reflected data as of the end fiscal year 2012. The report for fiscal year 2013 has not yet been submitted to Congress, and, therefore, the cost data for that year still reflects allocations or planned costs.

[9] During the preliminary assessment, the Corps uses available information, including a search of historical records, to determine whether the property was ever under the jurisdiction of DOD and owned or controlled by the United States, and if hazards caused by DOD's use

may be present. If the Corps determines that the property was under the jurisdiction of DOD and owned or controlled by the United States, but it does not find evidence of any hazards caused by DOD, it designates the property as NDAI. If, however, the Corps determines that hazards caused by DOD prior to October 17, 1986, may be present, then the Corps begins further study.

End Notes for Appendix III

[1] Interior uses the term "record" to refer to the information documented in the department's EDL database module.

[2] In 2005, Interior's Office of Inspector General recommended that the Office of Environmental Policy and Compliance develop a management information system that captures consistent and accurate data for all hazardous materials sites. To implement this recommendation, the Office of Environmental Policy and Compliance created the Environmental Management Information System. The EDL module of the database was implemented in the third quarter of fiscal year 2006 and is used by the department in preparing its quarterly and annual financial statements.

[3] Interior defines environmental assessment or due care as the process followed by a bureau or office to use reasonable effort to examine a location of concern—an area on department lands that is suspected to be contaminated based on known past activities or observed and reported physical indicators—to identify the presence or likely presence of contamination at concentrations significant enough to require further study or cleanup.

[4] According to Interior's *Environmental and Disposal Liabilities Identification, Documentation and Reporting Handbook v 3.0,* groundwater and surface water treatment actions to restore water quality to a protected level are considered part of the remedial action for the first 10 years of operation and operation and maintenance for any additional years.

In: Potentially Contaminated Sites …
Editor: Roger L. Brooks
ISBN: 978-1-63483-076-8

Chapter 2

COMPREHENSIVE ENVIRONMENTAL RESPONSE, COMPENSATION, AND LIABILITY ACT: A SUMMARY OF SUPERFUND CLEANUP AUTHORITIES AND RELATED PROVISIONS OF THE ACT*

David M. Bearden

SUMMARY

Congress enacted the Comprehensive Environmental Response, Compensation, and Liability Act of 1980 (CERCLA; P.L. 96-510) in response to a growing desire for the federal government to ensure the cleanup of the nation's most contaminated sites to protect the public from potential harm. The Superfund Amendments and Reauthorization Act of 1986 (P.L. 99-499, SARA) clarified the applicability of the statute's requirements to federal facilities, and modified various response, liability, and enforcement provisions. Several other laws also have amended CERCLA for specific purposes, including relief from cleanup liability for certain categories of parties, and the

* This is an edited, reformatted and augmented version of the Congressional Research Service Publication, CRS Report for Congress R41039, dated June 14, 2012.

authorization of federal assistance for the cleanup of abandoned or idled "brownfields" where the presence or perception of contamination may impede economic redevelopment.

CERCLA authorizes cleanup and enforcement actions to respond to actual or threatened releases of hazardous substances into the environment, but generally excludes releases of petroleum and certain other materials covered by other federal laws. Considering the limitation of federal resources to address the many contaminated sites across the United States, CERCLA directs the Environmental Protection Agency (EPA) to maintain a National Priorities List (NPL) to identify the most hazardous sites for the purpose of prioritizing cleanup actions. The states and the public may participate in federal cleanup decisions at NPL sites. The states primarily are responsible for pursuing the cleanup of sites not listed on the NPL, with the federal role at these sites limited mainly to addressing emergency situations.

CERCLA established a broad liability scheme that holds past and current owners and operators of facilities from which a release occurs financially responsible for cleanup costs, natural resource damages, and the costs of federal public health studies. At waste disposal sites, generators of the wastes and transporters of the wastes who selected the site for disposal also are liable under CERCLA. The liability of these "potentially responsible parties" (PRPs) has been interpreted by the courts to be strict, joint and several, and retroactive. At contaminated federal facilities, federal agencies are subject to liability under CERCLA as the owners and operators of those facilities on behalf of the United States. Federal agencies also may be liable in instances in which an agency generated or transported waste for disposal at a non-federal facility.

CERCLA established the Hazardous Substance Superfund Trust Fund to pay for the cleanup of sites where the PRPs cannot be found or cannot pay. A combination of special taxes on industry and general taxpayer revenues originally financed the Superfund Trust Fund, but the authority to collect the industry taxes expired on December 31, 1995. Over time, Congress increased the contribution of general revenues to make up for the shortfall from the expired industry taxes. General revenues now provide most of the funding for the trust fund, but other monies continue to contribute some revenues (i.e., cost-recoveries from PRPs, fines and penalties for violations of cleanup requirements, and interest on the trust fund balance). The availability of these trust fund monies under the Superfund program is subject to appropriations by Congress. Private settlement funds deposited into site-specific Special

Accounts within the Superfund Trust Fund also are available to EPA, but are not subject to discretionary appropriations.

Considering the liability of the federal government at its own facilities, the cleanup of federal facilities is not funded with Superfund Trust Fund monies under the Superfund program, but with other federal monies appropriated to the agencies responsible for administering the facilities. However, EPA and the states remain responsible for overseeing and enforcing the implementation of CERCLA at federal facilities to ensure that applicable cleanup requirements are met.

INTRODUCTION

By the end of the 1970s, Congress had enacted several environmental laws to regulate sources of pollution in the United States, but had not yet addressed responsibility for contamination resulting from releases of pollutants into the environment. In the late 1970s, the discovery of severely contaminated sites, such as "Love Canal" in New York and Times Beach in Missouri, raised questions as to whether there should be a federal role in cleaning up environmental contamination to protect the public from potential harm. Congress enacted the Comprehensive Environmental Response, Compensation, and Liability Act of 1980 (CERCLA; P.L. 96-510) to authorize the federal government to clean up contaminated sites in the United States and to make the "potentially responsible parties" connected to those sites financially liable for the cleanup costs. CERCLA created the Superfund program to carry out these authorities. The Environmental Protection Agency (EPA) administers the program. Subsequent amendments to CERCLA also authorized EPA to administer a separate grant program to support the cleanup of abandoned or idled "brownfields" properties to encourage their redevelopment.

CERCLA established a broad liability scheme that holds both past and current owners and operators of contaminated facilities financially responsible for the costs of cleanup. At waste disposal sites, generators of the waste sent to the site for disposal, and transporters of the waste who selected the site for disposal, also are responsible for the cleanup costs. If these potentially responsible parties cannot be found or cannot pay for the cleanup, CERCLA authorizes the federal government to finance the cleanup to ensure the protection of human health and the environment. These costs borne by the federal government are referred to as "orphan shares." The broad liability scheme of CERCLA is intended to capture all parties that may have had some involvement in the actions that resulted in contamination of the environment, in order to minimize the burden of the costs of cleanup on the general taxpayer who had no involvement. This approach to liability is based on the principle that polluters should be required to pay for the environmental damage that they cause, often referred to as the "polluter pays principle."

CERCLA established the Hazardous Substance Superfund Trust Fund to finance cleanup actions taken by the federal government at contaminated sites where the potentially responsible parties cannot pay or cannot be found. A combination of special taxes on industry and revenues from the General Fund of the U.S. Treasury initially financed the Superfund Trust Fund, but the authority to collect the industry taxes expired at the end of 1995. As the remaining revenues were expended over time, Congress increased the contribution of general Treasury revenues in an effort to make up for the shortfall from the expired industry taxes. The availability of Superfund Trust Fund monies to finance the cleanup of contaminated sites is subject to appropriations by Congress.

Considering the liability of the federal government as a potentially responsible party at its own facilities, the cleanup of federal facilities is not funded with Superfund Trust Fund monies under the Superfund program, but with other federal monies appropriated for other programs administered by the agencies responsible for these facilities. The Department of Defense (DOD) and the Department of Energy (DOE) administer the cleanup of most contaminated federal facilities. EPA and the states are responsible for overseeing and enforcing the implementation of CERCLA at federal facilities to ensure that applicable requirements are met.

To prioritize cleanup actions, CERCLA directed EPA to establish and maintain a National Priorities List (NPL) of the most contaminated sites in the United States which present the greatest risks to human health and the environment. The NPL includes both non-federal sites and federal facilities that are deemed to present a sufficient level of risk to warrant listing. EPA may require the potentially responsible parties to directly perform or pay for cleanup actions themselves. Alternatively, EPA may clean up a contaminated site up-front with appropriated Superfund monies and later recover those funds from the potentially responsible parties (with the exception of the cleanup of federal facilities which must be funded up-front by the administering agencies). In the event that the potentially responsible parties cannot pay or cannot be found, appropriated Superfund monies may be used to pay the orphan shares of cleanup costs at a site, under a cost-sharing agreement with the state in which the site is located.

The following sections of this report summarize the major cleanup authorities of CERCLA and other relevant provisions of the act. The topics discussed herein include the overall scope and reach of these statutory authorities, the process under which cleanup actions are selected and carried out at individual sites, the financial liability of potentially responsible parties for the costs of cleanup actions, the Superfund Trust Fund that may pay for cleanup actions when the potentially responsible parties cannot pay or cannot be found, enforcement of cleanup liability against the potentially responsible parties to minimize the need for federal tax revenues to finance the cleanup of contaminated sites, the applicability of CERCLA to federal facilities, and federal assistance for the cleanup of brownfields properties.

It should be emphasized that how and to what degree a specific contaminant at an individual site must be cleaned up under CERCLA are not specified in the law itself. The specific actions that are required to clean up contaminants at individual sites are determined on a site-by-site basis. Although CERCLA established a general process for making cleanup decisions, more specific direction is provided in EPA regulation and agency guidance. Other federal agencies that administer the cleanup of federal facilities under CERCLA have developed additional guidance documents that apply to their own respective facilities. Although the statutory authorities upon which federal agencies have based their cleanup regulations and guidance are discussed in this report, the content of these regulations and guidance is not examined here.

As such, this report summarizes selected statutory provisions of CERLCA, but does not discuss agency regulations and guidance that may provide more detailed direction for carrying out cleanup actions at individual sites.

MAJOR AMENDMENTS

Congress has amended CERCLA on numerous occasions to clarify the applicability of the cleanup authorities of the statute, and to provide relief from liability for certain categories of parties who may not have been involved in actions that led to contamination, or who may have contributed only certain quantities or types of waste to a site. Congress also has amended the statute to authorize federal assistance for the cleanup of abandoned or idled "brownfields" properties to encourage their redevelopment. Further, certain amendments have addressed unique cleanup challenges at federal facilities, such as the cleanup of unexploded ordnance on decommissioned military training ranges in the United States, and responsibility for the cleanup of contaminated federal property when it is transferred out of federal ownership.

The Superfund Amendments and Reauthorization Act of 1986 (SARA; P.L. 99-499) clarified that federal facilities are subject to the cleanup requirements of CERCLA to the same extent as non-federal entities, and amended various response, liability, and enforcement provisions of the law. The 1986 amendments also renewed the authorization of appropriations for EPA's Superfund program through FY1991, and established a separate Defense Environmental Restoration Program within the Department of Defense (DOD) to address contamination at active and decommissioned military facilities in the United States.

Title VI of the Omnibus Budget Reconciliation Act of 1990 (P.L. 101-508) extended the authorization of appropriations for EPA's Superfund program through FY1994, and Title XI of that statute extended the authority to collect the special Superfund taxes on industry through December 31, 1995. Although reauthorizing legislation has been introduced in various Congresses, the taxing authority for the Superfund Trust Fund has not been renewed to date, nor has the authorization of appropriations for EPA's Superfund program been extended. Instead, Congress has continued to fund the Superfund program primarily with general Treasury revenues through the annual appropriations process. Congress has annually authorized and appropriated funding for the Defense Environmental Restoration Program each year since its establishment. Most of this funding is supported with general Treasury revenues, with the exception of some revenues generated from the sale or lease of closed military bases which help fund their cleanup.

In 1992, the Community Environmental Response Facilitation Act (P.L. 102-426) amended the federal facility provisions of CERCLA to facilitate the transfer of uncontaminated parcels of surplus federal property on which hazardous substances or petroleum products were not released. Section 334 of the National Defense Authorization Act for FY1997 (P.L. 104-201) further amended CERCLA to allow the transfer of contaminated surplus federal property before cleanup is complete, if assurances are provided to guarantee that the property will be cleaned up to a level that would be suitable for its intended use after transfer.

Other amendments have attempted to address the fairness of the liability scheme of CERCLA, either by limiting or eliminating the liability of certain categories of parties. In 1996, the Asset Conservation, Lender Liability, and Deposit Insurance Protection Act (Subtitle E, Title II, Division A of P.L. 104-208) amended CERCLA to protect certain fiduciaries and financial lenders from liability. In 1999, the Superfund Recycling Equity Act (Title VI, Appendix I of P.L. 106-113) exempted generators and transporters of recyclable scrap materials from cleanup liability under CERCLA, if the person who received the materials disposed of them instead and the disposal resulted in contamination. There had been some concern that the potential liability of generators and transporters under CERCLA could be a deterrent to recycling.

In 2002, the Small Business Liability Relief and Brownfields Revitalization Act (P.L. 107-118) provided relief from cleanup liability for (1) persons who contributed very small quantities of waste or only municipal solid (i.e. non-hazardous) waste to a site, (2) owners of property that became contaminated merely as a result of migration from a contiguous property owned by another person, and (3) "bona fide" prospective purchasers who otherwise may be hesitant to acquire a contaminated property because of potential cleanup liability once acquiring ownership. The 2002 act also established more specific criteria for exempting "innocent" owners of contaminated property from cleanup liability, if they purchased the property without knowledge of the existing contamination and they had no involvement in actions that led to contamination.

As required by the statute, persons seeking an exemption from liability as a "bona fide" prospective purchaser, contiguous property owner, or "innocent" landowner must have performed "all appropriate inquiry" into the prior uses of the property *before* acquiring ownership, and must take "reasonable steps" *after* acquiring ownership to prevent potentially harmful exposure to environmental contamination. Because of these requirements, such persons still may bear some responsibility for managing contamination on their properties, even though they may be exempt from liability for more extensive cleanup actions that may be taken under CERCLA.

In addition to providing relief from liability for certain categories of parties, P.L. 107-118 authorized federal grants to assist in the cleanup of "brownfields" properties. Brownfields properties typically are abandoned, underutilized, or idled sites where the known or suspected presence of contamination, and the potential for cleanup liability, could be viewed as a deterrent to purchase the property for redevelopment. Brownfields properties tend to be less contaminated than sites listed on the NPL, but may need some cleanup to make them suitable for reuse. EPA originally had established a program in 1993 to provide federal assistance for the cleanup of brownfields properties using the general cleanup authorities of CERCLA as the legal basis for this assistance. P.L. 107-118 provided explicit statutory authority for this purpose, and established a separate Brownfields grant program within EPA, apart from the Superfund program.

Table 1 lists CERCLA as enacted in 1980 and the major amendments to the law noted above.

FEDERAL RESPONSE AUTHORITIES

Section 104(a) of CERCLA specifically authorizes the President to respond to a release (or substantial threat of a release) of a hazardous substance into the environment, or of a pollutant or contaminant which may present an "imminent and substantial danger to the public health or welfare."[1] As authorized by Section 115 of CERCLA,[2] the President delegated the response authorities of CERCLA to EPA and other federal agencies by Executive Order.[3] EPA may respond to releases on the land, and the U.S. Coast Guard may respond to releases into inland river ports and harbors, the Great Lakes, and U.S. coastal waters. If a release were to occur at a federal facility, the agency that administers that facility is authorized to take response actions, subject to oversight and enforcement by EPA and the states in which those facilities are located. Federal funding to carry out response actions under CERCLA is subject to appropriations by Congress.

Notification of a release of a hazardous substance is the action that may trigger a federal response under CERCLA. Section 103(a) requires the party responsible for a release to notify the National Response Center if the quantity of the release exceeds the regulatory limit established for that particular substance.[4] These limits are referred to as "reportable quantities," which are specified in federal regulation.[5] State or local officials, or members of the public, who observe or suspect a release of a hazardous substance also may report the incident. Once a release is reported, the National Response Center is to notify the appropriate federal agency that would be responsible for carrying out the President's response authorities under Section 104(a), and for taking any federal enforcement actions that may be necessary against the parties responsible for the release.

Response actions taken under CERCLA most often entail cleanup activities involving the containment, removal, or treatment of environmental contamination to prevent potentially harmful exposure, but may include the temporary or permanent relocation of potentially exposed individuals if warranted. Congress has excluded certain types of environmental contamination from the response authorities of CERCLA, which may be addressed under other federal environmental laws. These exclusions are provided within the statutory definitions of key terms upon which the response authorities of CERCLA hinge, including the terms "hazardous substance," "pollutant or contaminant," and "release." In addition to these exclusions, Congress has placed general limitations on the extent to which response actions may be taken under CERCLA to address releases of hazardous substances, pollutants, or contaminants in certain situations. In effect, these exclusions and limitations may restrict the applicability or scope of the response authorities of CERCLA at a particular contaminated site.

Table 1. Comprehensive Environmental Response, Compensation, and Liability Act (CERCLA) and Major Amendments to the Statute

Year	Title of Statute	Public Law Number
1980	Comprehensive Environmental Response, Compensation, and Liability Act of 1980	P.L. 96-510
1986	Superfund Amendments and Reauthorization Act of 1986	P.L. 99-499
1990	Omnibus Budget Reconciliation Act of 1990	P.L. 101-508, Title VI, § 6301, Title XI, Subtitle B, Part IV, § 11231
1992	Community Environmental Response Facilitation Act	P.L. 102-426
1996	Asset Conservation, Lender Liability, and Deposit Insurance Protection Act	P.L. 104-208, Division A, Title II, Subtitle E
1996	National Defense Authorization Act for Fiscal Year 1997	P.L. 104-201, § 334
1999	Superfund Recycling Equity Act	P.L. 106-113, Appendix I, Title VI
2002	Small Business Liability Relief and Brownfields Revitalization Act	P.L. 107-118

Petroleum Exclusion

The response authorities of CERCLA do not extend to releases of petroleum. Section 101(14) of CERCLA generally excludes releases of petroleum, including crude oil and any fraction thereof, from the definition of a "hazardous substance" for the purposes of the statute.[6] Section 101(33) does the same for the definition of "pollutant or contaminant."[7] Petroleum releases are covered instead by other statutes. The Oil Pollution Act of 1990 (P.L. 101-380) is the primary federal law that addresses releases of petroleum.[8] Other federal laws also provide authorities to respond to petroleum releases in specific situations. For example, Section 311(c) of the Clean Water Act authorizes the federal actions to respond to releases of petroleum into or on the navigable waters of the United States and adjoining shorelines.[9] Section 9003(h) of the Solid Waste Disposal Act provides federal response authorities for petroleum leaked from underground tanks.[10] In practice, CERCLA has been applied to the cleanup of some wastes containing petroleum only if the wastes also contained hazardous substances that were not part of the petroleum product itself.[11]

Other Exclusions

Section 101(22) of CERCLA also excludes certain types of releases from the definition of the term "release," thereby removing such releases from the statute's reach.[12] A specific category of nuclear materials is excluded from the definition of release, including "source, byproduct, or special nuclear material" released from a nuclear incident or at certain processing sites. The disposal and cleanup of these materials are subject to the Atomic Energy Act. With the exception of these specific nuclear materials, CERCLA generally applies to the release of radionuclides. In federal regulation, EPA has designated several hundred radionuclides as hazardous substances that are subject to the authorities of CERCLA.[13] Section 101(22) also excludes three other types of releases from the response authorities of CERCLA: (1) a release that would result in exposure solely within the workplace; (2) emissions from engine exhaust of a motor vehicle, train, aircraft, vessel, or power pumping station; and (3) the "normal" application of fertilizer. There also are certain situations identified in CERCLA in which a party would not be subject to liability, such as the proper application of a registered pesticide product[14] or a federally permitted release of a hazardous substance.[15] However, response authority under the statute generally would remain available to EPA in these two instances, just not the enforcement of liability.

Limitations on Response Actions

Section 104(a)(3) limits the extent to which actions may be taken under CERCLA to respond to releases of hazardous substances, pollutants, or contaminants in certain situations. Response actions generally may not be taken in situations involving (1) releases of naturally occurring substances in their unaltered form; (2) releases from products (such as asbestos) that are part of a residential, business, or community structure or building; or (3) releases into public or private drinking water supplies due to deterioration of supply systems through ordinary use.[16] However, in the event of a public health or environmental emergency declared by the President, CERCLA authorizes response actions to be taken under the statute in any of these three situations, if no other person has the authority and capability to respond in a timely manner.

Prioritization and Procedures for Response Actions

Section 105(a) of CERCLA required the President to develop a National Hazardous Substance Response Plan to establish procedures and standards for prioritizing and responding to releases of hazardous substances, pollutants, and contaminants into the environment. The law directed the President to incorporate these procedures and standards into the National Oil and Hazardous Substances Pollution Contingency Plan (referred to as the National Contingency Plan for short, or NCP). As delegated by the President, EPA promulgated the National Hazardous Substance Response Plan in federal regulation as part of the NCP.[17] These regulations govern any response actions taken under CERCLA.

Consistent with the purpose of the NCP, Section 105(a) of CERCLA also required the President to develop a National Priorities List (NPL) of the most hazardous sites in the United States as an administrative mechanism to prioritize response actions.[18] The President has delegated this task to EPA. The NPL must be updated at least once annually. Section 105(c) primarily requires the use of a Hazard Ranking System (HRS) to determine which sites warrant placement on the NPL.[19] The system scores each site based on certain factors, such as the quantity and nature of hazardous substances; the likelihood of the migration of contamination in groundwater, surface water, and air; and the proximity to human populations and sensitive environments. Because of this range of factors, the severity of contamination alone may not necessarily be sufficient cause to list a site on the NPL. For example, a geographically isolated site with substantial contamination still may not score highly enough on the HRS to warrant placement on the NPL, if the distance from human populations prevents the likelihood of exposure.

In addition to the use of the HRS to evaluate eligibility for listing a site on the NPL, there are two other mechanisms under which EPA also may list a site. First, Section 105(a) allowed each state the one-time opportunity to designate a single site within its borders as the state's highest priority for listing on the NPL. Second, EPA may list a site for which the Agency for Toxic Substances and Disease Registry (ATSDR) has issued a public health advisory, if EPA also determines that the contamination presents a significant public health threat and that its use of "remedial" authority will be more cost-effective than its sole use of "removal" authority without listing the site.[20] As discussed below in the "Scope of Response Actions" section, a site must be listed on the NPL as a condition for the availability of Superfund appropriations to perform remedial actions, but removal actions are not subject to such condition.

EPA has listed over 1,600 sites on the NPL over time, including federal facilities. EPA has deleted over 300 of these sites once EPA determined, in concurrence with the states, that the long-term cleanup objectives had been met. The vast majority of the sites were listed based on EPA's evaluation of the potential risks using the HRS, but some sites have been listed as a result of states designating them as their top priority and as a result of an ATSDR public health advisory.[21]

Scope of Response Actions

CERCLA authorizes two types of response actions: "removal" and "remedial" actions. These terms are defined in Sections 101(23)[22] and 101(24)[23] of CERCLA respectively. Removal does not necessarily mean the physical removal of contamination from the soil, surface water, or groundwater, and remedial actions do not necessarily involve treatment of contamination. Rather, both actions may involve various methods to prevent exposure to contamination, including the relocation of potentially exposed individuals if warranted. It should be noted that the NCP allows remedial actions to be financed with Superfund monies only at sites listed on the NPL, whereas removal actions may be financed with Superfund monies at non-NPL sites to address emergency situations.[24] This restriction is intended to reserve Superfund monies for costlier remedial actions at NPL sites that are thought to present the greatest risks. This funding restriction in the regulations is based on the statutory requirement of Section 105(a) of CERCLA for EPA to prioritize contaminated sites for the purpose of taking remedial actions.

Removal actions tend to be shorter term actions that address more immediate risks, whereas remedial actions tend to be longer term actions that offer a more permanent solution. As such, remedial actions often entail more extensive and costly measures. Because of the typically greater extent and cost of remedial actions, they are subject to more in-depth review in the form of a Remedial Investigation and Feasibility Study (RI/FS). An RI/FS involves an investigation of the contamination to assess potential risks of exposure and a study of the feasibility of remedial alternatives to address those risks. Remedial actions also are subject to public participation requirements under Section 117 of CERCLA.[25] (See the "Public Participation" section of this report.) Removal actions are not subject to a similar degree of review or public comment because of the perceived need for swifter response to address more immediate risks.

Section 104(c)(1) generally restricts the timing of removal actions funded with Superfund monies to one year and the cost to $2 million, with exceptions provided in certain situations.[26] For example, a remedial action may exceed these limitations if the continuance of the removal action would contribute to the remedial action planned at the site. These general timing and cost limitations on removal actions are intended to ensure that removal actions are not pursued on a broader scale as a way to avoid the more in-depth review required of remedial actions.

However, CERCLA does not impose these limitations on a removal action funded by a responsible party with its own funds, nor by a federal agency at a federal facility with dedicated monies appropriated to that agency for that purpose apart from Superfund. From a practical standpoint, imposing the above timing and cost limitations on removal actions at many federal facilities administered by the Department of Defense and Department of Energy could constrain the needed scope of removal actions, as cleanup challenges are often greater at these federal facilities in comparison to non-federal sites.

Federal-State Cost Sharing

Section 104(c)(3) of CERCLA requires the state in which a non-federal NPL site is located to agree to share the costs of remedial actions at that site, as a condition of obligating federal Superfund monies to finance those actions.[27] States are not responsible for sharing the costs of cleanup at sites where the potentially responsible parties pay for the cleanup, including federal facilities that are funded by the federal agencies that administer them. Rather, the federal government and the states are to share the costs of assuming the responsibility for the orphan shares of the cleanup costs, for which there are no viable parties to pursue.

This cost-sharing requirement in Section 104(c)(3) is intended to reduce the financial burden on the federal taxpayer presented by the often long-term financial commitment involved in carrying out a remedial action. Notably, CERCLA does not require states to agree to share the costs of removal actions, which typically are less costly as a result of their smaller scope. Consequently, federal Superfund monies may be used to finance the entire costs of removal actions.

At a site where the state must agree to share the costs of remedial actions as a condition of the obligation of federal Superfund monies, the state first must provide certain assurances of its financial commitments, specified in a binding contract or cooperative agreement with the federal government. Absent such contract or agreement, federal Superfund monies are not available to finance remedial actions at that site. To allow the obligation of federal Superfund monies to commence the remedial actions, the state must agree to pay 10% of the costs of those actions. If the site was owned or operated by the state, or a political subdivision of the state, at the time of disposal, the state must agree to pay at least 50% of the costs of the remedial actions.

In addition to the above conditions, the state must agree to perform future maintenance of the remedial actions for their expected operational life. The point of maintenance usually occurs after any necessary construction is complete and the remedial action is operating as intended. CERCLA authorizes a delay in the state's responsibility for the maintenance of groundwater or surface water remedies. Section 104(c)(6) allows a state to delay its maintenance responsibilities for the first 10 years of the operation of such remedial actions.[28] The statute allows a delay in the state's maintenance responsibility specifically for these types of actions to reduce the burden of those costs on the state, as the cleanup of groundwater or surface water tends to be more costly than other types of remediation. During the initial 10-year period, federal Superfund monies instead can be used to pay the maintenance costs of groundwater or surface water remedies.

SELECTION OF RESPONSE ACTIONS

Section 121(a) of CERCLA generally requires response actions at contaminated sites to achieve acceptable levels of exposure that would be protective of human health and the environment.[29] Response actions also are to be cost-effective over both the short-term and long-term, including the operation and maintenance of the action. Section 121(b) states a preference for the selection of remedial actions that involve treatment to "permanently and significantly" reduce the "volume, toxicity or mobility" of contamination, as opposed to actions that do not involve such treatment.[30]

Actions not involving treatment often entail the containment of wastes on-site, or the removal and disposal of wastes off-site. The containment of wastes on-site could present lingering health and environmental risks if the containment method were to fail over time. If the remedial action would result in wastes being left on-site, Section 121(c) requires the President to review the performance of the remedial action every five years to determine whether that action continues to be protective of human health and the environment.[31] If the action is not functioning as intended, the President may take additional remedial actions at the site to achieve the cleanup goal.

Although Section 121 includes certain requirements to govern the selection of remedial actions, it does not specify how clean an individual site must be to protect human health and the environment. Section 121 also does not identify the specific nature of the remedial actions that would be required to attain a cleanup goal at an individual site. Instead, these cleanup decisions are made on a site-by-site basis taking many factors into consideration, including the potential for human exposure based on the anticipated land use, and the technical and economic feasibility of cleanup alternatives to prevent exposure.

Cleanup Standards

The level of cleanup that is required can vary widely from site to site depending on the contaminants present, the cleanup standards or criteria that apply to those contaminants, and the response actions selected to attain those standards or criteria. Rather than specify standards or criteria for individual hazardous substances, Section 121(d) of CERCLA broadly requires that cleanup comply with applicable, relevant, and appropriate requirements (ARARs) to protect human health and the environment.[32] ARARs can include a host of federal or state standards, requirements, or other criteria. In this sense, CERCLA functions as an "umbrella" statute under which other statutes or regulations also may be applied to the cleanup of a contaminated site.

Section 121(d)(4) authorizes the waiver of a particular standard, if:

- the contemplated response action would be part of a larger remedial action that would meet the standard once the larger action is completed;
- compliance with the standard would result in a greater risk than the alternatives;
- compliance with the standard would be technically impracticable from an engineering perspective;
- an equivalent standard of performance would be attained;
- in the case of a state standard, the state has not consistently applied that standard elsewhere within its jurisdiction; or
- meeting the standard would not provide a balance between the need for protection of public health and welfare and the environment at the site under consideration, and the availability of monies in the Superfund Trust Fund to respond to more immediate risks at other sites.[33]

Although CERCLA generally does not list specific standards that may apply to the cleanup of an individual site, there are two sets of standards cited in Section 121(d) that broadly apply to the selection of remedial actions at any site. First, the law requires remedial actions to achieve a level of cleanup that would attain Maximum Contaminant Levels (MCLs) established for current or potential sources of drinking water under the Safe Drinking Water Act.[34] Second, remedial actions must be consistent with other water quality criteria established under Sections 303[35] or 304[36] of the Clean Water Act. However, the applicability of these sets of standards to an individual site remains limited to circumstances in which the standards still are deemed "relevant and appropriate," consistent with the underlying premise of an ARAR.

State Participation

CERCLA authorizes a broad role for states to participate in the cleanup process. States must agree to share in the costs of remedial actions at non-federal NPL sites as a condition of the obligation of federal Superfund monies. In acknowledgment of their sharing of the costs of cleanup, Section 121(f) of CERCLA requires that states be afforded opportunities for "substantial and meaningful involvement" in initiating, developing, and selecting remedial actions.[37] However, there are certain limitations on the involvement of states in cleanup decisions at federal facilities, as states do not share in the costs of cleanup at these facilities. If a state wishes to challenge a remedial decision of a federal agency at a facility which that agency administers, Section 121(f)(3) requires that the state show that the decision of the agency is not supported by "substantial evidence."[38]

Public Participation

CERCLA also provides a role for the general public in commenting on the selection of remedial actions at individual sites. This role is similar to that under many other federal laws that require the opportunity for the public to comment on certain types of federal decisions. Section 117 of CERCLA requires EPA, or other federal agency responsible for administering and funding the cleanup of a contaminated site, to provide the public an opportunity to comment on proposals for the selection of remedial actions.[39] Once a final decision is made, public notice of the decision must be provided, with an explanation of any "significant" differences from the proposed action and a response to each "significant" public comment on the proposed action.

The opportunity for public comment required by Section 117 of CERCLA applies only to decisions on remedial actions. Decisions on removal actions are not subject to these requirements because of the presumed need for expedited action to address more immediate risks. In practice, EPA and other federal agencies typically notify the public of the selection of removal actions to inform communities of the nature and timing of such actions. To assist the public in understanding technical information presented in cleanup decision documents, Section 117(e) of CERCLA authorizes technical assistance grants of up to $50,000 for community groups.[40] These grants are available only to affected communities at sites listed on the NPL.

AGENCY FOR TOXIC SUBSTANCES AND DISEASE REGISTRY

Section 104(i) of CERCLA established the Agency for Toxic Substances and Disease Registry (ATSDR) primarily to assess potential health risks at NPL sites.[41] The ATSDR assesses individual sites based on the likelihood of human exposure to contamination through the air, soil, surface water, groundwater, and other pathways such as consumption of contaminated food sources. The purpose of these assessments is two-fold: to inform the public of potential health hazards at a contaminated site, and to aid decision-makers in evaluating what cleanup actions may be warranted to prevent potentially harmful exposure. Although the findings of the ATSDR may be used to inform the selection of cleanup actions, the agency does not have any authority to dictate cleanup decisions. In addition to site-specific assessments, Section 104(i) directs the ATSDR to prepare toxicological profiles of hazardous substances commonly found at NPL sites to identify potential health effects that can result from exposure.

Section 104(i) of CERCLA also authorizes the ATSDR to carry out several other functions intended to protect public health. For example, the agency is authorized to provide medical care and testing to individuals in the event of a public health emergency caused by, or believed to be caused by, exposure to toxic substances. CERCLA does not provide any criteria as to what constitutes a public health emergency for this purpose, presumably leaving the declaration of such an emergency to the discretion of the ATSDR. As with other roles, the resources of the agency to fulfill this role are subject to appropriations by Congress. To date, the ATSDR has not used its authority under CERCLA to declare a public health emergency. In practice, the agency's role has focused on educating the public about known health risks from exposure to hazardous substances, and assessing potential risks at individual sites to aid in informing cleanup decisions.

FINANCIAL LIABILITY

Section 107 of CERCLA identifies the categories of potentially responsible parties connected with a contaminated site who are liable for the costs of response actions that EPA deems necessary to protect human health and the environment.[42] Such parties also are liable for damages for injury to, destruction of, or loss of natural resources resulting from a release of a hazardous substance, including the costs of assessing such injury, destruction, or loss; and the costs of public health assessments carried out by the ATSDR under Section 104(i) of CERCLA. The following sections discuss the categories of parties who are liable under Section 107 of CERCLA, the reach of liability, defenses to liability, and limitations on the liability of certain categories of parties.

Categories of Potentially Responsible Parties

Section 107(a) identifies four categories of potentially responsible parties who are liable for the costs of response actions, natural resource damages, and public health assessments associated with the release or threatened release of a hazardous substance:

- any person who currently owns or operates a facility or vessel from which a hazardous substance was released;
- any person who at the time of disposal of a hazardous substance owned or operated the facility at which such disposal occurred;
- any person who arranged for the disposal or treatment of a hazardous substance (often referred to as a generator of waste), and any person who arranged for the transport of a hazardous substance for disposal or treatment; and
- any person who accepts or accepted a hazardous substance for transport to a disposal or treatment facility, incineration vessel, or site selected by such person.[43]

In the context of liability, it should be noted that financial responsibility for cleanup costs may extend to actions beyond a facility boundary, if a hazardous substance were to migrate (i.e., move or spread) through the environment. Section 101(8) of CERCLA defines the term "environment" to include not only the land, but also surface water, groundwater, or ambient air.[44] Consequently, cleanup actions may be necessary not only on the facility where the initial release occurred, but anywhere the hazardous substance may migrate through the environment. For example, hazardous substances that migrate into groundwater or surface water can travel some distance, even miles, and can necessitate cleanup actions across a larger area than where the release first occurred.

Reach of Liability

Over time, the courts have interpreted liability under Section 107 of CERCLA to be strict, joint and several, and retroactive. This judicial interpretation is rooted in case law, legislative history, and the definition of liability in Section 101(32) of CERCLA[45] that applies the same standards of liability as in Section 311 of the Clean Water Act.[46]

- Strict liability means that a party can be held liable regardless of whether the conduct of that party was negligent.
- Joint and several liability means that one or more of the liable parties can be held responsible for the full cost of the cleanup at a site, regardless of the degree of involvement in the contamination. However, Section 113(f)(1) of CERCLA allows a party to seek recovery of some of its cleanup costs from other parties at a site through contribution claims in court.[47] In deciding such claims, a court is to base the allocation of cleanup costs on "equitable factors." In the event that a party can show that the waste *it* sent to the site could not have contributed to the contamination, joint and several liability is not to apply to that party.
- Retroactive liability means that parties are liable for the cleanup of hazardous substances released prior to the enactment of CERCLA on December 11, 1980. However, Section 107(f)(1) extends liability for natural resource damages only to releases that occurred on or after the enactment of CERCLA, which resulted in injury to, destruction of, or loss of the natural resources.[48]

It should be emphasized that the above description of the basic liability standards of CERCLA merely offers a brief summary of the broad reach of the statute, as generally interpreted by the courts over time. As such, this description does not examine the complexities of individual court decisions on these matters. Since the enactment of CERCLA in 1980, well over one thousand court decisions have interpreted these basic liability standards under the statute to determine the financial responsibility of potentially responsible parties for the costs of cleanup. How a court may view the cleanup liability of an individual party at any one site would depend on numerous legal issues that are beyond the scope of the summary of CERCLA offered in this report.

Defenses to Liability

Section 107(b) of CERCLA provides defenses to liability under certain circumstances.[49] A party cannot be held liable for the release or threatened release of a hazardous substance, and resulting injury to, destruction of, or loss of natural resources, if that party can provide evidence that the release or threatened release was caused solely by

- an act of God;
- an act of war;
- an act or omission of a third party with whom the defendant has no contractual relationship, if the defendant exercised due care with respect to the hazardous substance and took precautions against foreseeable acts or omissions of that third party and against the foreseeable consequences of such acts or omissions; or
- any combination of these three circumstances.

The third party defense sometimes is characterized as the "innocent" landowner defense, in the sense that it typically pertains to property owners who had no involvement in the actions that led to the contamination. Section 101(35) of CERCLA defines the term contractual relationship for the purpose of the third party defense, and specifies the conditions that a landowner must satisfy to claim the lack of a contractual relationship connecting the owner to the contamination.[50] See the "Bona Fide Prospective Purchasers and Innocent Landowners" section of this report below.

Limitations on Liability

To address the fairness of the liability scheme of CERCLA, Congress has amended the statute at various times to limit, or in some cases eliminate, the liability of certain categories of parties who may not have been involved in actions that resulted in contamination, who may have contributed only very small quantities or less toxic wastes to a contaminated site, or whose conduct Congress did not wish to discourage. These categories of parties include:

- response action contractors who merely perform the work to clean up a contaminated site, but who did not cause or otherwise contribute to the contamination;

- state and local governments that acquired contaminated property involuntarily through bankruptcy, tax delinquency, abandonment, or other circumstances, and did not cause or otherwise contribute to the contamination;
- persons who only hold a contaminated property in a fiduciary capacity;
- financial lenders who acquire financial interests or ownership of a contaminated property through foreclosure;
- generators and transporters of scrap materials intended for recycling, but instead may have been disposed of by other persons;
- persons who contributed only very small quantities of waste or only municipal solid (i.e. non-hazardous) waste to a site;
- service station dealers who only disposed of recycled oil that was not contaminated with hazardous substances, and who fully complied with federal regulations for managing the recycled oil;
- "innocent" landowners who purchased a property without knowledge of existing contamination, with respect to the third party defense noted above;
- other "innocent" owners of property that became contaminated merely through migration from a contiguous property where the initial release occurred; and
- "bona fide" prospective purchasers who otherwise may be hesitant to acquire a property on which contamination is known or suspected to be present, because of the potential liability for cleanup upon acquiring ownership.

Amendments to CERCLA that provided such limitations on cleanup liability for specific categories of parties are examined further below. As discussed in the "Other Exclusions" section of this report, there also are certain situations identified in CERCLA in which a party would not be subject to liability, such as the proper application of a registered pesticide product[51] or a federally permitted release of a hazardous substance.[52]

Cleanup Contractors

Soon after the enactment of CERCLA in 1980, it was realized that a private contractor hired to clean up a contaminated site may be exposed to potential liability as an operator of that site, or as a person who arranged for disposal or transport of waste in instances in which the contractor removed waste as part of the cleanup. This exposure to potential liability was viewed as a deterrent to private contractors being willing to clean up contaminated sites. In response to this concern, Section 119 of the Superfund Amendments and Reauthorization Act of 1986 added Section 119 to CERCLA to limit the liability of "response action" contractors who are hired to perform cleanup actions.[53]

Section 119(a)(1) states that cleanup contractors shall not be liable under CERCLA, or any other federal law, to any person for "injuries, costs, damages, expenses, or other liability" resulting from the release or threatened release of a hazardous substance, pollutant, or contaminant.[54] (However, no immunity from liability under state law is conferred under CERCLA.) Section 119(a)(2) states that a cleanup contractor shall not be exempt from federal liability for a release caused by that contractor as a result of conduct that is "negligent, grossly negligent, or which constitutes intentional misconduct."[55] Under certain circumstances, Section 119(c) authorizes the President the discretion to indemnify a cleanup contractor for negligent conduct, but not grossly negligent conduct or intentional misconduct.[56] Such indemnification is intended to cover a contractor's liability that cannot be covered by insurance at a "fair and reasonable" price.

Fiduciaries and Financial Lenders

Enacted in the 104th Congress, the Asset Conservation, Lender Liability, and Deposit Insurance Protection Act of 1996 (P.L. 104-208, Division A, Title II, Subtitle E of the Omnibus Consolidated Appropriations Act for FY1997) added Section 107(n) to CERCLA to limit the liability of persons who hold a facility or vessel only in a fiduciary capacity for another person to the value of the assets held on behalf of that person.[57] This limitation on liability is provided if the fiduciary did not cause or contribute to a release or threatened release of a hazardous substance from such facility or vessel. Fiduciaries who held an interest in a contaminated property had been concerned that joint and several liability under CERCLA could result in their financial liability exceeding the value of the assets held.

P.L. 104-208 also amended the definition of "owner or operator" in Section 101(20) of CERCLA[58] to exclude financial lenders that did not participate in the management of a facility or vessel from which there was a release or threatened release of a hazardous substance, but who held indicia of ownership primarily to protect security interests. Lenders also were exempted from liability as owners or operators of foreclosed properties, but only if they did not participate in the management of the facility or vessel prior to foreclosure. Lenders especially had been concerned about becoming liable for the cleanup of contaminated properties following foreclosure, when they become owners of the property, and hence could become liable for cleanup under Section 107.

Generators and Transporters of Recyclable Materials

Enacted in the 106th Congress, Title VI—Superfund Recycling Equity—of Appendix I of the Consolidated Appropriations Act for FY2000 (P.L. 106-113) added Section 127 to CERCLA to exempt certain parties involved in the recycling of scrap materials from cleanup liability as generators and transporters of wastes.[59] The exemption is available to persons who "arranged" for the recycling of scrap materials (by selling the materials or otherwise arranging for their recycling). Recyclers involved in these activities had been concerned about becoming liable as generators or transporters of wastes if they sold or transported scrap materials to a facility that disposed of the materials instead of recycling them as intended. Some had perceived this potential liability as a deterrent to recycling.

The exemption is available to the above persons only for materials that fall within the statutory definition of recyclable materials in CERCLA. Section 127(b) defines recyclable materials to include the following scrap materials: plastic, glass, textiles, rubber (other than whole tires), metal, or spent lead-acid, spent nickel-cadmium, and other spent batteries, and minor amounts of material incident to or adhering to the scrap material as a result of its normal and customary use prior to becoming scrap.[60] Two items are expressly excluded: (1) shipping containers of a certain capacity that contained a hazardous substance or onto which a hazardous substance adhered, and (2) materials containing polychlorinated biphenyls (PCBs) in excess of federal standards.

Even if a material can be considered recyclable within the above statutory definition, the exemption is not automatic. Section 127(c) requires the person who arranged for the recycling of the materials to demonstrate that certain criteria were met.[61] For example, the material must have been of commercial specification grade; a market must have existed for the material; a substantial portion of that type of material must have been made available for the manufacture of a new saleable product; and the material could have been used to replace, or to substitute for, virgin raw material. The person also must demonstrate that he or she exercised "reasonable care" to determine that the receiving facility where the materials were intended to be recycled was in compliance with federal, state, and local environmental laws.

Section 127(f) makes the exemption unavailable if the person claiming the exemption had reason to believe the scrap material would not be recycled by the receiving facility; that the material would be burned as fuel or for energy recovery or incineration; that the receiving facility was not in compliance with federal, state, and local environmental laws; or that hazardous substances had been added to the material; or if the person failed to exercise "reasonable care" in managing and handling the material.[62]

Contributors of "De micromis" and Municipal Solid Wastes

Enacted in the 107th Congress, Section 102(a) of the Small Business Liability Relief and Brownfields Revitalization Act (P.L. 107-118) amended Section 107 of CERLA to limit the liability of parties who contributed only certain quantities or types of wastes to sites listed on the NPL. The amendment did not extend these exemptions to such parties at sites not listed on the NPL. Section 102(a) of P.L. 107-118 added Section 107(o) to CERCLA to exempt from cleanup liability parties who generated or transported waste to a site listed on the NPL, if they contributed only "de micromis" amounts of hazardous substances to that site.[63] To qualify for this exemption, a party must demonstrate that it contributed less than 110 gallons of liquid materials or less than 200 pounds of solid materials containing hazardous substances. The exemption is not available to persons who contributed such quantities of wastes to a site on or after April 1, 2001.

Section 102(a) of P.L. 107-118 also added Section 107(p) to CERCLA to exempt from cleanup liability residential property owners, small businesses, and small non-profit organizations that contributed only municipal solid waste to a site listed on the NPL.[64] The exemption is limited to municipal solid waste generated by a household or that possesses characteristics typical of household waste. The exemption is available only to the generators of the waste and persons who arranged for the transport of the waste. The exemption is not available to the owner or operator of the disposal site where the waste was sent, such as a landfill, nor to transporters of the waste who selected the disposal site.

Bona Fide Prospective Purchasers and Innocent Landowners

Subtitle B of Title II of P.L. 107-118 authorized exemptions from cleanup liability for two categories of parties: (1) "bona fide" prospective purchasers and (2) owners whose properties became contaminated only as a result of migration from a contiguous property owned by another person. Subtitle B also established more specific criteria for the availability of the third party defense to "innocent" landowners who had no knowledge of existing contamination at the time of acquiring a property and had no involvement in the actions that led to the contamination. These exemptions are available to site owners who meet the requisite statutory criteria, regardless of whether the site is listed on the NPL. In contrast, the above exemptions from cleanup liability for contributors of de micromis amounts of wastes and municipal solid wastes are available only at sites listed on the NPL.

Prior to the enactment of P.L. 107-118, EPA had used its existing authorities under Section 122 of CERCLA[65] to enter into voluntary settlement agreements with prospective purchasers who had no involvement in the contamination, as a mechanism to limit their cleanup liability upon acquiring ownership of a contaminated property. EPA commonly referred to these agreements as "Prospective Purchaser Agreements."[66] This type of settlement usually capped a purchaser's financial responsibility for the cleanup, or required less extensive cleanup work by the purchaser. These agreements also typically included a covenant promising that the federal government would not sue for further liability, and offered protection from contribution claims by other liable parties. (See the "Voluntary Settlement Agreements" section of this report for additional discussion.) After the enactment of P.L. 107-118, a "bona fide" prospective purchaser who met the requisite statutory criteria could become eligible for an exemption from cleanup liability without entering into a formal settlement agreement with EPA.

Section 222 of P.L. 107-118 added Section 107(r) to CERCLA, exempting "bona fide" prospective purchasers of contaminated property acquired after the date of the enactment of P.L. 107-118 (January 11, 2002).[67] This exemption is not available to persons who purchased a contaminated property on or before January 11, 2002, and therefore is not retroactive. A person who knowingly purchased a contaminated property on or before that date must have entered into a Prospective Purchaser Agreement with EPA under Section 122 of CERCLA to limit his or her cleanup liability upon acquiring ownership. Section 222 of P.L. 107-118 also added Section 101(40) to CERCLA, defining the term "bona fide prospective purchaser" and specifying the criteria of eligibility for the exemption provided in Section 107(r) of the statute. [68]

In addition to "bona fide" prospective purchasers, Section 221 of P.L. 107-118 added Section 107(q) to CERCLA, exempting owners of contaminated property from cleanup liability if the contamination occurred only as a result of the migration of a hazardous substance from a contiguous property owned by another person. [69] To obtain the exemption, an owner of a contiguous property must have had no knowledge of the presence of the hazardous substance, nor the possibility of its migration, when acquiring ownership. Section 223 of P.L. 107-118 amended the definition of the term "contractual relationship" in Section 101(35) of CERCLA[70] to establish more specific criteria for "innocent" landowners to claim the third party defense against liability under Section 107(b)(3) of CERCLA.[71] Of importance, the exemptions for "bona fide" prospective purchasers and contiguous property owners reference these criteria in the definition of contractual relationship, making the criteria applicable to all three exemptions.

Under these criteria, an owner claiming an exemption as a "bona fide" prospective purchaser, "innocent" landowner, or contiguous property owner must have had no association with the activities that led to the contamination, and must have had no relationship with the persons who caused or contributed to the contamination (aside from a contractual relationship only involving the conveyance of the property). The key difference among these exemptions is that a "bona fide" prospective purchaser may know that a property is contaminated at the time of acquisition, and still be exempt from liability under CERCLA for the cleanup costs. A person claiming an exemption as an "innocent" landowner or a contiguous property owner must prove that he or she had no knowledge, or no reason to know, of the contamination at the time of acquisition.

To demonstrate that a best effort was made to know whether contamination may be present, CERCLA requires a person to have made "all appropriate inquiries" into the previous ownership and uses of the property.[72] A person seeking to claim an exemption from liability as a "bona fide" prospective purchaser, "innocent" landowner, or contiguous property owner must satisfy this requirement *before* acquiring ownership. As directed by P.L. 107-118, EPA promulgated regulations that identify specific measures a person must take to demonstrate that "all appropriate inquiries" were made satisfactorily.[73] The process outlined in the regulations for making "all appropriate inquiries" is similar to a preliminary site assessment, and must be performed by an environmental professional hired at the person's expense.

A person also must satisfy other conditions *after* receiving ownership of a property to maintain an exemption from liability under CERCLA as a "bona fide" prospective purchaser, "innocent" landowner, or contiguous property owner. The owner must take "reasonable steps" to (1) stop any continuing release of a hazardous substance; (2) prevent any future releases; and (3) prevent or limit exposure to any previously released hazardous substance.[74] The owner also must provide any legally required notices of the discovery of hazardous substances on the property, and must comply with any land use restrictions and institutional controls that may be put into place by regulators to prevent potential exposure to the hazardous substances. Satisfying these criteria can have the effect of minimizing, but not eliminating, an owner's responsibility for managing the contamination, even though the owner may be exempt from cleanup liability under CERCLA.

A person who discovered contamination upon conducting "all appropriate inquiries" must take "reasonable steps" to manage the contamination once becoming the owner to maintain exemption status as a "bona fide" prospective purchaser. If a person conducted "all appropriate inquiries" before acquiring a property and still did not discover the contamination, that person must take these steps once the existence of the contamination is known to maintain exemption status as an "innocent" landowner, or a contiguous property owner if the contamination resulted from migration. It should be emphasized that the burden of proof is on the person seeking an exemption from liability to demonstrate that "all appropriate inquiries" were made before acquiring ownership *and* that "reasonable steps" are taken after acquiring ownership to manage the contamination once its existence is known.

Hazardous Substance Superfund Trust Fund

CERCLA established the Hazardous Substance Superfund Trust Fund to provide a source of funds for the federal government to finance the cleanup of contaminated sites where the potentially responsible parties cannot pay or cannot be identified. This assumption of financial responsibility for these "orphan shares" of cleanup costs is intended to ensure that the actions necessary to protect human health and the environment are carried out. The availability of Superfund Trust Fund monies to pay for the cleanup of orphaned sites is subject to appropriations by Congress. Once appropriated, the availability of Superfund monies under EPA's Superfund program to pay for remedial actions is further subject to cost-sharing agreements with the states in which the sites are located, as discussed in the "State Participation" section of this report.

Original Taxing Authority

The special taxing authority to finance the Superfund Trust Fund expired at the end of 1995. Before this authority lapsed, three dedicated taxes on petroleum, chemical feedstocks (and imported chemical derivatives), and corporate income provided most of the revenues for the Superfund Trust Fund. Revenues from the General Fund of the U.S. Treasury also contributed to the trust fund to augment the dedicated taxes, but these general tax revenues were a relatively small portion of the total revenues to the trust fund during the time that the dedicated taxes were collected through the end of 1995.

As originally enacted in 1980, Section 211(a) of CERCLA authorized the Superfund excise taxes on petroleum[75] and chemical feedstocks.[76] Section 515(a) of the Superfund Amendments and Reauthorization Act of 1986 expanded the reach of the tax on domestically manufactured chemical feedstocks to include imported chemical derivatives.[77] Taxing imported derivatives was intended to compensate for the potential loss of revenues as overseas manufacturing of chemical feedstocks increased. Prior to expiration at the end of 1995, the Superfund excise tax on petroleum was 9.7 cents per barrel. The Superfund excise tax on chemical feedstocks and imported chemical derivatives varied from $0.22 per ton to $4.87 per ton, depending on the substance (with the exception of xylene which was taxed at a higher rate of $10.13 per ton in the initial years of the tax until 1992.) Section 516(a) of the Superfund Amendments and Reauthorization Act of 1986 established the special tax on corporate income to provide an additional revenue stream for the Superfund Trust Fund.[78] Prior to expiration in 1995, the Superfund tax on corporate income (formally referred to as the Corporate Environmental Income Tax) was 0.12% of corporate alternative minimum taxable income in excess of $2 million.

Whether to reinstate Superfund taxes has been a long-standing controversy since the taxing authority lapsed at the end of 1995. Congress has considered numerous bills to reauthorize the taxes, but none have been enacted to date. The reauthorization debate has centered around numerous "fairness" issues. Supporters of the taxes maintain that dedicated tax revenues for the Superfund program are necessary to ensure that polluters pay for the cleanup of contamination they have caused or may cause in the future, often referred to as the "polluter pays principle." In this sense, some have characterized Superfund taxes as an "insurance plan" for the public that is intended to provide resources for cleanup in the event that businesses may become bankrupt and cannot be pursued for their liability. On the other hand, opponents of the taxes have observed that not all of the individual businesses subject to the tax may have been involved in activities that resulted in contamination, and that the actual polluters are paying for the cleanup of most Superfund sites through enforcement actions under the liability provisions of CERCLA.

The extent to which Superfund taxes may have affected "innocent" businesses has been a principal question in the debate over the fairness of the tax structure. The Superfund tax on corporate income was intended to raise additional revenues from a wide range of businesses that may have benefitted from the use of hazardous substances in some way. However, this income tax captured all businesses that met the income threshold, regardless of whether a business may have used or disposed of any hazardous substances. Congress created the Superfund taxes on petroleum and chemical feedstocks based on the broadly held assumption that much of the environmental contamination in the United States had been caused as a result of industrial activities that involved these substances. However, not all petroleum and chemical companies may have been involved in actions that led to contamination.

The appropriateness of the Superfund tax on petroleum has been especially controversial in light of the exclusion of petroleum from the cleanup authorities of CERCLA. Because of this exclusion, monies from the Superfund Trust Fund generally have paid for the cleanup of petroleum contamination, only if the contamination includes hazardous substances that are not part of the petroleum product itself. Congress has established other trust funds to address releases of petroleum. Title V of the Superfund Amendments and Reauthorization of 1986 created the Leaking Underground Storage Tank Trust Fund to pay for actions to respond to petroleum released from underground tanks.[79] Title VIII of the Omnibus Budget Reconciliation Act of 1986 (P.L. 99-509) created the Oil Spill Liability Trust Fund to pay for actions to respond to surface releases of petroleum.[80]

Since Superfund taxing authority lapsed at the end of 1995, there have continued to be varying perspectives on how to fund the cleanup of contaminated sites in the most fair manner to ensure that the responsible parties satisfy their liability, while minimizing the financial burden on taxpayers who did not cause or otherwise contribute to the contamination, or who did not benefit in some way from the actions that resulted in the contamination. The current source of revenues for the Superfund Trust Fund is discussed below.

Current Source of Revenues

After the authority to collect the Superfund taxes expired, the remaining revenues from these taxes were expended by the end of FY2003, leaving revenues from the General Fund of the U.S. Treasury as the main source of monies for the Superfund Trust Fund. Although the Superfund taxes have expired, industry has continued to provide some of the funding for the trust fund via corporate income taxes that contribute to the General Fund. (Revenues to the General Fund consist of corporate income taxes, individual income taxes, and miscellaneous federal receipts and collections that are not dedicated to specific federal trust funds.)

In addition to general Treasury revenues, others sources of monies have continued to contribute some revenues to the Superfund Trust Fund for appropriation by Congress. Cleanup costs borne by the federal government that are later recouped from the potentially responsible parties are deposited into the trust fund (referred to as cost recoveries). These recouped funds can be made available for the cleanup of other sites where the potentially responsible parties cannot pay or cannot be found. Fines and penalties assessed against potentially responsible parties for violations of CERCLA are deposited into the trust fund as well. Interest also accrues on the trust fund balance. Collectively, these monies have been relatively small compared to the amount of general Treasury revenues that now support most of the annual discretionary appropriations from the trust fund to implement EPA's Superfund program. However, these other sources of monies do continue to help finance the trust fund, and to some extent reduce the need for general Treasury revenues at sites where the potentially responsible parties cannot be found or cannot pay.

Special Account Funds

Private settlement funds have been an additional source of monies for the Superfund Trust Fund. As amended in 1986, Section 122(b)(3) of CERCLA authorizes EPA to retain funds that it receives from private parties under voluntary settlement agreements to perform the cleanup of sites at which those parties may be liable.[81] (See the "Voluntary Settlement Agreements" section below.) These private settlement funds are deposited into site-specific Special Accounts within the Superfund Trust Fund, which are dedicated to the cleanup of the sites covered under the settlements. EPA has received nearly $4 billion in private settlement funds over time and has deposited these funds into over 1,000 Superfund Special Accounts since the establishment of the first account in FY1990.[82] These funds are available directly to EPA and are not subject to discretionary appropriations by Congress. Once all planned future work at a site is complete, EPA may "reclassify" the remaining balance of a Special Account for direct obligation to perform cleanup work at other sites, as a means to replace any appropriated funds that also may have been spent at the site covered by the Special Account. In other instances, EPA may transfer the remaining balance of a Special Account to the general portion of the Superfund Trust Fund, which would be subject to subsequent appropriation by Congress.

ENFORCEMENT MECHANISMS

There are three mechanisms through which the federal government can take actions to enforce cleanup liability under CERCLA, if the potentially responsible parties can be identified and have the financial capability to pay. These mechanisms include judicial or administrative orders, cost- recovery actions, and voluntary settlement agreements. Like the response authorities of CERCLA, these enforcement authorities are Presidential authorities. As discussed earlier in this report, a 1987 Executive Order delegated the President's response authorities under CERCLA to EPA and other federal agencies. This order also delegated the enforcement of the statute to EPA at sites on the land, and to the U.S. Coast Guard within inland river ports and harbors, the Great Lakes, and U.S. coastal waters. References in this report to the enforcement authorities of EPA apply equally to the U.S. Coast Guard within its delegated jurisdiction. CERCLA also authorizes citizen suits to enforce the cleanup requirements of CERCLA, but a cleanup action first must be completed before compliance with applicable requirements can be challenged. Each of these enforcement mechanisms is discussed below.

Judicial or Administrative Orders

Section 106(a) of CERCLA authorizes EPA to issue an administrative order, or to pursue a judicial order through the Department of Justice, to require a potentially responsible party to perform cleanup actions to address "an imminent and substantial endangerment to public health or welfare, or the environment" arising from an actual or threatened release of a hazardous substance.[83] Section 106(b)(1) authorizes fines of up to $25,000 per day for failure to comply with a cleanup order.[84] Section 107(c)(3) of CERCLA also allows a party that fails to comply with a cleanup order to be held liable for punitive damages up to three times the costs incurred by the United States out of the Superfund Trust Fund to carry out the cleanup action that the party did not perform.[85] Monies received by the United States for such punitive damages are to be deposited into the trust fund, and can be made available to finance the cleanup of other sites, subject to appropriations by Congress.

If the party who receives and complies with a Section 106 order can prove it is not liable under CERCLA, or that the cleanup actions required by EPA under the order were "arbitrary and capricious" or otherwise not in accordance with law, Section 106(b)(2) authorizes that party to be reimbursed from the Superfund Trust Fund.[86] This provision is intended to protect an innocent party from the costs of enforcement actions that may be imposed inappropriately upon that party, or to prevent a liable party from being required to pay for a more stringent cleanup than may be warranted to protect human health and the environment.

Cost-Recovery Actions

At some sites, EPA may spend Superfund Trust Fund monies upfront to initiate the cleanup if the potentially responsible parties are not yet identified, or if a cleanup order or settlement agreement with the identified parties is not yet finalized. In the event that EPA does expend Superfund monies at a site with viable parties, reimbursement may be included in the terms of any administrative settlement agreement that may be entered into with the parties. EPA also may pursue recovery of Superfund monies from the parties through judicial actions, in conjunction with the Department of Justice.

Section 107(a) of CERCLA specifically authorizes EPA to recover Superfund monies from the potentially responsible parties, as long as those actions are not inconsistent with the NCP.[87] States and Indian tribes, and any other person, who chooses to perform cleanup actions also may recover their costs from the potentially responsible parties, as long as those actions are consistent with the NCP. The costs of health effects studies carried out by the ATSDR under Section 104(i) of CERCLA,[88] and damages for injury to, destruction of, or loss of natural resources (and the assessment of such injury, destruction, or loss), are recoverable as well.

Section 113(g)(2) of CERCLA limits the time during which a cost-recovery action may be commenced against a potentially responsible party, which could reduce a party's financial liability at a site if recovery is not sought quickly enough.[89] Cost-recovery actions must be filed within three years after the completion of a removal action, except for removal actions allowed to extend beyond the general time limit of 12 months. For these lengthier removal actions, the costs can be sought within six years after the determination was made to extend the timing beyond 12 months. Cost-recovery actions must be commenced within six years after the initiation of the physical construction of a remedial action. If the remedial action is initiated within three years after the completion of the removal action that preceded it, the costs of that removal action may be recovered as part of the recovery of the costs of the remedial action that followed.

Voluntary Settlement Agreements

If a potentially responsible party is willing to resolve its liability voluntarily, Section 122 of CERCLA[90] gives EPA the discretion to enter into an administrative settlement agreement with that party instead of pursuing an enforcement action through a judicial or administrative order under Section 106, or a cost-recovery action under Section 107. Avoiding an enforcement action by EPA through a voluntary settlement agreement can save a party the costs of litigation, possibly motivating a party to agree to settle its liability.

Section 122(f) gives EPA the discretion to include a covenant in the agreement promising that the federal government will not sue for further liability.[91] Such a covenant can provide an incentive for a party to agree to perform specific cleanup actions or to make a monetary payment in exchange for a cap on its liability. As discussed in the "Special Account Funds" section above, Section 122(b)(3) authorizes EPA to retain the funds received under a settlement and directly use the funds to fulfill the terms of the settlement.[92] A party who voluntarily settles its liability at a site also is afforded protection from contribution claims by other parties at that site, under Section 113(f)(2) of CERCLA.[93] Such protection is intended to offer yet another incentive for a party to settle, especially if a contribution claim by another party appears imminent.

Whether to enter into a settlement agreement with EPA under Section 122 is entirely voluntary on the part of the potentially responsible party. However, once finalized, the terms of the agreement to perform specific cleanup actions or to make a monetary payment are binding on the party who entered into the agreement. If the party fails to perform the agreed-upon cleanup actions or to pay the agreed-upon costs of the cleanup, Section 109 of CERCLA authorizes civil penalties of up to $25,000 each day that the violation of the agreement continues to occur.[94]

Ability-to-Pay Considerations

In enforcing cleanup liability under CERCLA, EPA has the discretion to consider a potentially responsible party's financial capability in determining that party's share of the cleanup costs. A party with limited financial capability who desires to reduce its share of the cleanup costs may request a reduction in its share through the negotiation of a voluntary settlement agreement with EPA under Section 122 of CERCLA, discussed above. In the negotiation process, the party seeking the reduction must submit financial information to EPA for the agency's consideration to determine whether the party's ability to pay the cleanup costs may in fact be limited.

In 1995, EPA issued its first guidance document on ability-to-pay considerations for use in settlement negotiations. The agency revised its guidance again in 1997.[95] EPA formulated its guidance based on court interpretations of the reach and intent of the cleanup liability provisions of CERCLA,[96] and the agency's policy of balancing two fundamental interests: ensuring that a potentially responsible party satisfies its liability for cleanup, while at the same time not creating an undue financial hardship on that party, or on those who may be dependent upon that party.

In 2002, Congress included provisions in Section 102(b) of Title I of P.L. 107-118 that amended Section 122(g) of CERCLA to establish a new category of *de minimis* settlement that explicitly endorsed EPA's policy to reduce a party's share of the cleanup costs based on that party's ability to pay.[97] Prior to this amendment, CERCLA explicitly authorized *de minimis* settlements only for owners of property who were not involved in the release of hazardous substances and who had no knowledge of any hazardous substances on the property; and for persons who contributed a relatively small amount of hazardous substances that were minimally toxic in comparison to other hazardous substances at the site.

In determining whether a party satisfactorily demonstrates a limited ability to pay, EPA must consider the ability of the person to pay for cleanup actions and "... still maintain its basic business operations, including consideration of the overall financial condition of the person and demonstrable constraints on the ability of the person to raise revenues."[98] Consistent with earlier EPA guidance, the amendment explicitly requires a person seeking a reduced settlement to provide EPA with the financial information that would be necessary to determine the ability of that person to pay for cleanup actions at the site concerned.[99]

If EPA were to grant a reduced settlement, the person who is the subject of the settlement must waive all contribution claims against other potentially responsible parties at the site, unless EPA were to determine that requiring a waiver would be an "unjust" condition.[100] A reduced settlement does not remove a party from the responsibility to provide information and access to the site in the future that may be necessary to carry out the cleanup.[101] After a reduced settlement is finalized, EPA must notify any other potentially responsible parties at the site who have not resolved their liability with the federal government.[102]

Financial capability aside, EPA still may decline a potentially responsible party's request for a reduced settlement, if the agency determines that the party has failed to comply with any request for access, information, or an administrative subpoena in relation to the site, or has impeded or is impeding, through action or inaction, the performance of a cleanup action at the site.[103] If EPA were to determine that a potentially responsible party is not eligible for a reduced settlement, EPA is required to provide the reasons for the determination in writing to the potentially responsible party who requested the reduced settlement.[104]

EPA's determination of a party's eligibility for a reduced settlement is not subject to judicial review,[105] nor is a dispute over an ability-to-pay determination of the agency within the jurisdiction of the EPA Environmental Appeals Board.[106] Consequently, EPA would appear to have final authority to determine a party's ability to pay its share of the cleanup costs, based on the financial information submitted by that party in the settlement negotiation process. In practice, a party only can pay to the extent of its actual financial capability, to the point of bankruptcy. The authority of EPA to reduce a party's share of the cleanup costs is intended to avoid such financial outcomes as a consequence of cleanup liability under CERCLA.

Enforcement Discretion

CERCLA does not require EPA to use any one particular enforcement mechanism at an individual site, but allows the agency enforcement discretion to select which of the above mechanisms would be the most effective in achieving cleanup goals. EPA typically attempts to negotiate voluntary settlement agreements with the potentially responsible parties first, and usually turns to the use of Section 106 orders or Section 107 cost-recovery actions when a negotiated settlement appears unlikely. At a site where there are multiple potentially responsible parties, EPA also has the enforcement discretion to pursue the liability of all, some, or only one party. Even when enforcing against less than all parties, EPA still may recover the full amount of cleanup costs through joint and several liability. As described earlier in this report, joint and several liability means that any liable party can be held responsible for the full cost of cleanup, regardless of the degree of involvement.

EPA usually pursues the liability of parties at a site who are thought to have contributed more greatly to the contamination, and to be more capable of performing or paying for the cleanup. This selective approach is intended to reduce the enforcement transactions costs to the federal government. For the purpose of fairness, Section 113(f)(1) of CERCLA authorizes the parties who are enforced against to recover some of their costs from other potentially responsible parties whom EPA did not elect to pursue.[107] Section 113(f)(3)(B) also authorizes parties who have resolved all (or some) of their liability under settlements with EPA to seek contribution from other parties who are not participants in the settlements.[108] Section 113(f)(2) explicitly protects parties from contribution claims who have entered into settlements with EPA to resolve their liability under Section 122 of CERCLA.[109]

Citizen Suits

Although EPA is responsible for enforcing cleanup liability, Section 206 of the Superfund Amendments and Reauthorization Act of 1986 added Section 310 to CERCLA authorizing citizens to challenge the adequacy of a cleanup action in court.[110] The timing of a citizen suit for these purposes is limited. Section 113(h)(4) of CERCLA does not permit a citizen suit to be brought for violation of a cleanup requirement until the selected cleanup action at a site is completed.[111] Further, a citizen suit may not be brought with regard to a removal action at a site where a remedial action is planned. This limitation on the timing of a citizen suit is intended to allow the complete implementation of cleanup actions planned at a site, prior to subjecting the adequacy of those actions to judicial review to assess their compliance with CERCLA.

Once the cleanup actions are completed, Section 310(a)(1) authorizes a citizen to commence a civil action against any person who is alleged still to be in violation of a "standard, regulation, condition, requirement, or order,"[112] including any provision of a federal facility cleanup agreement issued under Section 120.[113] Section 310(b)(1) requires such suits to be brought in the district court for the district in which the violation is alleged to have occurred.[114] Section 310(c) authorizes the court to require actions to correct the violation and impose civil penalties.[115] Section 310(d)(1) requires the plaintiff to notify the President, the state in which the violation is alleged to have occurred, and the alleged violator 60 days in advance of commencing a civil action,[116] in a manner prescribed by federal regulation.[117] Section 310(d)(2) prohibits citizen suits if the President already has commenced and is "diligently prosecuting" an enforcement action against the potentially responsible party.[118]

Section 310(a)(2) also authorizes a citizen to commence a civil action against the President or any other officer of the United States, including the Administrators of EPA and ATSDR, for alleged failure to perform any non-discretionary act or duty required under CERCLA, including such act or duty required at a federal facility.[119] Section 310(b)(2) requires such suits to be brought in the United States District Court for the District of Columbia.[120] Section 310(c) authorizes the court to order the President or other officer of the United States to perform the act or duty concerned.[121] Section 310(e) requires the plaintiff to notify the Administrator of EPA, or other department or agency, 60 days in advance of commencing a civil action for the alleged failure to perform a non- discretionary act or duty,[122] in a manner prescribed by federal regulation.[123]

Citizen suits are not available for alleged failure of the President or an officer of the United States to perform a non-discretionary act or duty under Section 311 of CERCLA.[124] This provision states that the Secretary of Health and Human Services "shall" establish and support a research and training program to enhance understanding of the potential health risks associated with exposure to hazardous substances. The program also is to research methods and technologies that would detect hazardous substances in the environment and reduce their amount and toxicity. The program is to be carried out through the awarding of grants, cooperative agreements, and contracts, the funding for which is subject to annual appropriations by Congress.

FEDERAL FACILITIES

After CERCLA was enacted in 1980, questions arose as to whether Congress intended federal facilities to be subject to the cleanup authorities and liability provisions of the statute to the same extent as non-federal facilities. As originally enacted, Section 101(21) of CERCLA defined the term "person" for the purposes of the statute to include the federal government,[125] meaning that the reference to persons who may be held liable under Section 107 may include the federal government. However, the original enactment of the law did not otherwise explicitly address the liability of federal agencies, nor the applicability of other provisions of the statute to federal agencies. Section 120 of the Superfund Amendments and Reauthorization Act of 1986 added Section 120 to CERCLA to clarify that federal departments and agencies are subject to the requirements of CERCLA to the same extent as other entities, including the liability and enforcement provisions of the law.[126] To comply with CERCLA, the federal agency with administrative jurisdiction over a federal facility is responsible for performing and paying for the cleanup of contamination out of its own budget, subject to appropriations by Congress.

Section 111(e) of CERCLA explicitly prohibits the use of Superfund Trust Fund monies to clean up federal facilities, as these monies are dedicated to paying for the cleanup of sites where the potentially responsible parties cannot be identified or cannot pay.[127] However, Section 111(e)(3) does allow the use of Superfund Trust Fund monies at an individual federal facility to provide alternative water supplies, if groundwater contamination has migrated beyond the boundary of that facility, and there are other potentially responsible parties connected to that facility in addition to the United States.[128] In all other instances, Superfund Trust Fund monies are not available for the cleanup of federal facilities.

Congress appropriates funding to various accounts of federal agencies to pay for the performance of the cleanup of federal facilities. These funds generally are intended to fulfill the liability of the United States as the owner or operator of these facilities.[129] However, these accounts do not constitute a cleanup liability fund in a broader sense. The funds are authorized to pay for the performance of the cleanup of the federal government's own facilities by federal agencies. However, the funds are not explicitly authorized to pay cleanup cost-recovery or contribution claims that may be submitted to the United States by other parties, either at federal facilities or at non-federal sites where a federal agency may be held liable as a generator or transporter of wastes sent to a site for disposal. The Judgment Fund of the U.S. Treasury has been the source of payments for cleanup claims submitted to the United States to satisfy the federal share of liability under CERCLA, and compromise settlements for such claims.[130] By statute, the Judgment Fund is a permanent, indefinite appropriation that is intended to pay monetary claims against the United States, which are not otherwise provided by Congress through separate appropriations.[131]

EPA and the states play a role in overseeing and enforcing the implementation of CERCLA at federal facilities, although the agencies that administer these facilities actually fund their cleanup. Section 120(e) of the law requires EPA to take the lead in overseeing the cleanup of federal facilities listed on the NPL,[132] but Section 120(f) allows states and local governments to participate in cleanup decisions.[133] The states play a more prominent role in overseeing the cleanup of federal facilities not listed on the NPL. While CERCLA authorizes EPA and the states to oversee the cleanup of federal facilities, certain provisions of the law can limit their ability to direct or dictate how the cleanup process may be carried out.

As discussed below, CERCLA gives EPA decision-making authority to select remedial actions at federal facilities listed on the NPL, but does not explicitly authorize EPA to direct the schedule of performing those actions, nor how those actions are to be operated and maintained over the long- term to ensure their performance. Further, EPA's enforcement of cleanup requirements at federal facilities through court actions is complicated by the limited ability of one federal agency to sue another. With respect to states, CERCLA requires the opportunity to be involved in cleanup decisions, but does not give states any decision-making authority. In practice, these limitations may restrict the extent to which EPA and the states may oversee the cleanup of federal facilities, even though Section 120 of CERCLA specifically requires federal facilities to comply with the requirements of the statute to the same extent as other entities.

Facilities on the National Priorities List

Within 6 months of the listing of a federal facility on the NPL, Section 120(e)(1) of CERCLA requires the federal agency with administrative jurisdiction over the facility to consult with EPA and the appropriate state authorities to begin a Remedial Investigation/Feasibility Study (RI/FS).[134] As discussed earlier in the "Scope of Response Actions" section of this report, an RI/FS involves an investigation of contamination to assess potential risks to human health and the environment, and a study of the feasibility of the remedial alternatives to address those risks. While consultation with EPA and state authorities is required, CERCLA does not give explicit decision-making authority to EPA or the states to dictate precisely how a federal agency performs this investigation and study phase of the cleanup process.

Within 180 days of the completion of the RI/FS and review by EPA, Section 120(e)(2) requires the federal agency with administrative jurisdiction over the facility to enter into an interagency agreement with EPA to govern the remedial actions to be taken at that facility.[135] This agreement provides an opportunity for EPA to formalize how the other federal agency will carry out the cleanup of the facility to satisfy the requirements of CERCLA. Section 120(e)(4) identifies four elements that are to be included in each interagency agreement: (1) a list of the remedial alternatives considered at the facility, (2) identification of the remedial actions selected from among the alternatives, (3) a schedule for completing each remedial action, and (4) arrangements for any long-term operation and maintenance activities that may be necessary to ensure the performance of the remedial actions over time.[136]

If EPA and the federal agency with administrative jurisdiction over the facility cannot agree on the selection of the remedial actions in negotiating an interagency agreement, Section 120(e)(4)(A) authorizes the Administrator of EPA to resolve the dispute and select the remedial actions he or she deems most appropriate to protect human health and the environment.[137] Although the Administrator may delegate this dispute-resolution authority to an officer or employee of EPA, Section 120(g) prohibits the transfer of the Administrator's authorities under Section 120 to any other agency, official, or employee of the United States, by Executive Order of the President or otherwise, or to any other person.[138] This prohibition primarily is intended to ensure that the role of EPA is maintained in determining the selection of remedial actions.

CERCLA does not provide the Administrator of EPA decision-making authority with respect to other elements of an interagency agreement for a federal facility listed on the NPL, namely the schedule for completing the remedial actions and arrangement for any long-term operation and maintenance activities that may be necessary to ensure the performance of those actions over time. These latter elements would appear to be subject to negotiation between EPA and the federal agency with administrative jurisdiction over the facility. If consensus cannot be reached and the agreement finalized within the statutory deadline of 180 days from the completion of the RI/FS, Section 120(e)(5) requires the federal agency with administrative jurisdiction over the facility to report the delay to Congress.[139]

With respect to the timing of the cleanup, Section 120(e)(3) requires the federal agency responsible for the facility to complete the remedial actions "as expeditiously as practicable" once those actions are selected, but does not indicate a specific time frame or deadline for their completion.[140] The timing of a remedial action ultimately depends on the technical feasibility of that action and the availability of appropriations by Congress. Accordingly, Section 120(e)(3) requires federal agencies to notify Congress of the amount of funding needed to carry out the selected remedial actions at their facilities in their annual budget requests.

Notably, the lack of a final interagency agreement governing an entire facility does not preclude individual remedial actions from proceeding to address discrete contaminated sites at a facility. Further, removal actions intended to address more immediate risks are not subject to an interagency agreement. The main reason for this difference is that the time required to finalize an agreement may delay a removal action needed to address an emergency situation. Because of these reasons, some cleanup actions may proceed without an interagency agreement in place, in effect leaving EPA with less formal means to oversee the cleanup.

States and local governments also may play a role in the cleanup of federal facilities listed on the NPL. Section 120(f) of CERCLA authorizes states and local governments to participate in the planning and selection of remedial actions at federal facilities.[141] Participation by states and local governments is to include, but is not limited to, review of all applicable data as it becomes available, and the development of studies, reports, and plans. Section 120(f) specifies that the opportunity for state officials to participate in cleanup decisions at federal facilities is to be provided in accordance with Section 121(f).

As discussed earlier in the "State Participation" section of this report, Section 121(f) requires states to be afforded opportunities for "substantial and meaningful involvement" in initiating, developing, and selecting remedial actions.[142] Section 121(d) also allows state standards to be applied to a remedial action, thereby offering additional opportunity for state participation.[143] However, Section 121(f)(3) specifies that, if a state wishes to challenge a remedial decision at a federal facility, that state must show that the decision is not supported by "substantial evidence" to compel the selection of a different remedy at that facility.[144] Unlike Section 120(f), the participation requirements of Section 121(f) are not extended to local governments.

Facilities Not on the National Priorities List

States play a more prominent role in overseeing the cleanup of federal facilities not listed on the NPL. In acknowledgement of this role, Section 120(a)(4) of CERCLA clarifies the reach of state law at contaminated federal facilities that are not listed on the NPL.[145] This provision stipulates that state cleanup standards or requirements shall apply to a federal facility that is not on the NPL only to the same extent as those standards or requirements would apply to a non-federal site located in that state. In practical terms, a state may not require more stringent cleanup at a federal facility than it would require at a non-federal site possessing comparable characteristics and conditions under which exposure to contamination may occur.

While state cleanup laws generally can be applied to federal facilities not listed on the NPL, CERCLA does not require federal agencies to enter into formal agreements with states to govern cleanup requirements in a fashion similar to interagency agreements with EPA. However, states may have other authorities to identify and enforce cleanup requirements at federal facilities that they oversee. Most notably, federal facilities that store, treat, or dispose of hazardous waste are subject to permits issued by states with federal authority delegated under the Solid Waste Disposal Act.[146] These permits can require "corrective action" to clean up contamination that may result from waste management or disposal practices.[147] In contemplation of such situations, Section 120(i) of CERCLA states that nothing in CERCLA may affect or impair the obligation of federal agencies to comply with requirements of the Solid Waste Disposal Act at the facilities that they administer, specifically including corrective action requirements.[148]

A corrective action that a state may require under the Solid Waste Disposal Act can be similar in scope to a removal or remedial action under

CERCLA. This similarity can result in essentially the same stringency of cleanup in practice, regardless of which statute is applied. As such, Solid Waste Disposal Act permits at federal facilities not on the NPL can function much like CERCLA interagency agreements at federal facilities on the NPL, specifying individual actions required to clean up contamination. In this sense, Solid Waste Disposal Act permits can provide a means for a state to formalize and enforce cleanup requirements at many federal facilities that are not listed on the NPL, for which an interagency agreement with EPA is not required.

Transfer of Contaminated Federal Property

Section 120(h) of CERCLA generally requires the United States to clean up contaminated federal property prior to transferring the property out of federal ownership.[149] The policy premise of this provision is that the United States should assume full responsibility for the cleanup of contamination caused by federal activities, and not shift the burden of that responsibility to the recipient merely as a consequence of acquiring the property. Section 120(h) applies to all contaminated federal property declared surplus to the needs of the federal government. The agency with administrative jurisdiction over a surplus federal property usually performs and pays for the cleanup of contamination to fulfill the financial liability of the United States.

As is the case with federal facilities that remain in federal ownership, funds available for the cleanup of surplus federal properties are subject to appropriations by Congress, and are not eligible for Superfund monies. For example, the Department of Defense performs and pays for the cleanup of surplus federal property on closed military installations out of funds appropriated to the Base Realignment and Closure (BRAC) accounts.

Section 120(h) does not bind the United States to cleaning up a surplus federal property for any one particular use. As a result, the reuse of a property is negotiated between the administering federal agency and the recipient of the property. Disagreements over reuse can arise if the recipient intends to use the property for a purpose that would necessitate a level of cleanup that the federal agency may consider too costly, relative to available appropriations to fund the cleanup. The capabilities of cleanup technologies also could constrain the reuse of a surplus federal property, if it would be impractical to achieve a level of cleanup that would be needed to make the property suitable for a use desired by the recipient.

Continuing Liability of the United States

Consistent with the policy premise of Section 120(h) and retroactive liability under Section 107, the United States remains responsible for contamination found not to have been sufficiently remediated after the property is transferred out of federal ownership. Section 120(h)(3) requires the continuing liability of the United States to be specified through a "covenant" incorporated into the deed transferring the property out of federal ownership.[150] The covenant must warrant that all remedial actions necessary to protect human health and the environment have been taken before the date of transfer, and that the United States shall conduct any additional remedial actions found to be necessary after the date of transfer. A clause also must be included in the deed granting the United States access to the property to perform cleanup actions for which it may be responsible.

In practice, the contents of a deed can place certain limitations on the continuing responsibility of the United States. A deed to a transferred federal property typically warrants cleanup only to a level suitable for the land use negotiated prior to transfer. In some cases, a deed may include a restriction prohibiting certain uses that would be considered unsuitable relative to the level of cleanup performed by the United States. Under such deed restrictions, the United States typically assumes responsibility for additional cleanup only to the extent that more work is found to be needed to make the originally agreed-upon use suitable.

If the new owner later wishes to use the property for a different purpose, the new owner typically must assume responsibility for the additional cleanup costs to make the property suitable for that purpose. In some instances, a deed may prohibit certain land uses even if the new owner is willing to pay the cleanup costs. For example, a deed to a decommissioned military training range may prohibit residential or other uses because of the limitations of cleanup technologies to detect and remove unexploded ordnance. Cleanup capabilities may be constrained especially when ordnance is located beneath the surface, or concealed on the surface by dense vegetation.

Transfer of Uncontaminated Parcels

Some surplus federal properties may contain a mix of contaminated and uncontaminated parcels of land. Although the clean parcels may be ready for reuse, the requirement to clean up the contaminated parcels under Section 120(h) of CERCLA prior to transfer could delay the conveyance of the property as a whole. To address such situations, the 102nd Congress enacted the Community Environmental Response Facilitation Act (CERFA; P.L. 102-426) in 1992. This law amended Section 120(h) by adding a new subsection (4) that authorizes the transfer of uncontaminated parcels on a surplus federal property while cleanup continues on the contaminated parcels.[151] This parcel-by-parcel approach is intended to avoid potential delays in the transfer of "clean" surplus federal lands for reuse, especially such lands on closed military installations where economic redevelopment is desired to replace lost jobs. In the event that previously unknown contamination is later discovered after transfer out of federal ownership, Section 120(h)(4)(D) requires that a deed to an uncontaminated parcel still include a covenant warranting that the United States shall conduct any cleanup actions that may become necessary.[152]

Early Transfer of Contaminated Parcels

The cleanup of a contaminated parcel may take several years or more, depending on the type and level of contamination, technical feasibility of cleanup actions, and availability of appropriations to pay for the cleanup. In such situations, the requirement to complete cleanup prior to transfer out of federal ownership could result in delaying the transfer. Enacted in the 104th Congress, Section 334 of the National Defense Authorization Act for FY1997 (P.L. 104-201) amended Section 120(h)(3) of CERCLA to add a new subsection (C) that allows the transfer of a contaminated parcel on a surplus federal property before cleanup is complete, if certain conditions are satisfied.[153] Although Congress enacted this amendment in annual defense authorization legislation, this authority applies to any surplus federal property administered by any federal agency, not just surplus U.S. military property.

Section 120(h)(3)(C) specifically authorizes a deferral of the cleanup covenant to allow the transfer of title to a contaminated parcel on a surplus federal property before cleanup is complete. Federal agencies often refer to this deferral of the covenant as an "early" transfer, although the statute does not use this term. The deed to a contaminated property transferred out of federal ownership must contain assurances that the cleanup still will be carried out after the property leaves federal ownership. The federal agency responsible for the performance of the cleanup also must identify the funding needed to carry out the cleanup in its annual budget requests.

The deed also must restrict the use of the property to purposes that would be protective of human health and environment, while the cleanup proceeds. For example, at the time of transfer, a property may be suitable for industrial use because the risks of exposure to contamination may be within an acceptable range, whereas other uses that would result in potentially harmful exposure would be restricted until the property is cleaned up sufficiently for that purpose. Once cleanup is complete, the United States remains obligated to provide a covenant at that time, warranting that all necessary actions to protect human health and the environment have been taken to make the property suitable for its intended, eventual use.

The early transfer of a contaminated surplus federal property that is listed on the NPL is subject to the concurrence of the Administrator of EPA and the governor of the state in which the facility is located. The early transfer of a contaminated surplus federal property that is not listed on the NPL still requires the concurrence of the governor of the state in which the facility is located, but not EPA. Federal agencies proposing an early transfer also must provide the public at least 30 days advance notice and an opportunity to comment on the proposed transfer before it is executed.

Cleanup Authorities Specific to Military Facilities

Considering that U.S. military facilities constitute a substantial portion of the inventory of contaminated federal facilities, Section 211 of the Superfund Amendments and Reauthorization Act of 1986 required the Secretary of Defense to establish the Defense Environmental Restoration Program to perform the cleanup of U.S. military facilities.[154] This provision also authorized dedicated Defense Environmental Restoration appropriations accounts to fund the program. Section 211 requires the Secretary of Defense to perform the cleanup of U.S. military facilities under the program in accordance with Section 120 of CERCLA, which in turn specifies the applicability of all of the requirements of CERCLA and the liability and enforcement provisions of the law. Section 211 also requires the Secretary to consult with EPA in implementing the Defense Environmental Restoration Program. Notably, the provisions of Section 211 did not amend CERCLA itself, but were treated as "stand-alone" provisions that apply strictly to U.S. military facilities.[155]

The scope of the Defense Environmental Restoration Program includes the performance of the cleanup of military facilities in the United States that *are* or *were* under the jurisdiction of the Secretary and owned by, leased to, or otherwise possessed by the United States at the time the contamination occurred, and as such may include both active and decommissioned military facilities.[156] The inclusion of decommissioned facilities within the program's scope is consistent with retroactive liability under Section 107 of CERCLA, under which the Department of Defense can be held liable for cleanup as the past owner and operator of those facilities. The scope of the program also includes the correction of other environmental damage that may present an imminent and substantial endangerment to the public health, welfare, or the environment (such as the presence of unexploded ordnance on decommissioned military training ranges), and the demolition and removal of unsafe buildings and structures for safety purposes. The scope of the program does not include the payment of cleanup cost-recovery or contribution claims that may be submitted to the United States by other parties to satisfy federal liability arising from activities of the Department of Defense. As discussed above, the Judgment Fund has been the source of federal monies for the payment of such cleanup claims.

The Defense Environmental Restoration Program initially focused on the cleanup of hazardous substances without a consolidated effort in place to address the safety risks of unexploded ordnance on decommissioned military training ranges. In response to concerns among the public about these potential safety hazards, the 107th Congress included provisions in Sections 311[157] and 312[158] of the National Defense Authorization Act for FY2002 (P.L. 107-107) that expanded the scope of the Defense Environmental Restoration Program to include the cleanup of unexploded ordnance, discarded military munitions, and munitions constituents (i.e., hazardous substances leached from munitions into the environment) on decommissioned military training ranges and munitions disposal sites in the United States. The Department of Defense established the Military Munitions Response Program as a sub-element within the Defense Environmental Restoration Program to carry out these requirements.

The statutory authority of the Military Munitions Response Program extends only to decommissioned military training ranges and munitions disposal sites in the United States, but not to operational ranges.[159] Since the enactment of specific cleanup authorities for military facilities in the 1986 amendments to CERCLA, the Department of Defense has expressed long-standing concern that the carrying out of cleanup actions on an operational range could prevent or interrupt its active use for training, and thereby possibly impair military readiness. So far, operational ranges have been subject to federal waste disposal regulations promulgated under the Solid Waste Disposal Act, but not cleanup under CERCLA unless the contamination migrates off-range. EPA promulgated these disposal regulations in 1997, referred to as the Military Munitions Rule.[160] Under this rule, munitions on an operational range are not considered hazardous waste, and therefore are not subject to hazardous waste disposal requirements under the Solid Waste Disposal Act, until they are removed from the range. Upon removal, their disposal is subject to permit requirements for hazardous waste disposal.

Munitions typically are removed from an operational range only to the extent necessary to ensure safe access by military personnel for training purposes. Consequently, much of the munitions may remain on an operational range indefinitely, unless contamination from munitions were to migrate off-range and present a risk of exposure. In such situations, removal of munitions could be pursued to eliminate the source of the contamination. Absent off-range migration, munitions generally may be left on a range as long as the range remains in operational status. In such circumstances, cleanup of the munitions typically is not required until the range is closed, and the range then becomes eligible for cleanup under the Military Munitions Response Program.

National Security Exemption

Although Section 120 of CERCLA clarified the applicability of the statute to federal facilities, Section 120(j) authorized the President to exempt an individual federal facility from a requirement of CERCLA on a case-by-case basis if the exemption would be necessary to protect national security.[161] This exemption is intended to prevent situations in which a federal facility may become unavailable for purposes essential to protecting national security, if carrying out a specific cleanup action somehow may interfere with those purposes. Section 120(j) specifically authorizes the President to exempt an individual facility administered by the Department of Defense or the Department of Energy from compliance with a requirement of CERCLA, if the President deems such an exemption necessary to protect national security.

The President must notify Congress within 30 days of the issuance of an exemption and explain the reason for it. The time period of an exemption initially is limited to one year, but the President may renew it annually with notification to Congress. To date, a national security exemption under CERCLA has not been invoked at any facility of the Department of Defense or the Department of Energy. Instead, contaminated facilities of both departments have been made subject to the cleanup requirements of CERCLA.

BROWNFIELDS PROPERTIES

In 1993, EPA established an element within the Superfund program to assist communities with the cleanup of certain lower risk sites that did not warrant placement on the NPL, but at which cleanup was desired to encourage economic redevelopment. The purpose of the program was to provide federal financial assistance for the cleanup of properties referred to as "brownfields." These properties typically are abandoned, idled, or underutilized, and on which known or suspected contamination is perceived as a deterrent to redevelopment by prospective purchasers who may be hesitant about becoming liable for cleanup once acquiring ownership.

EPA initially used Superfund appropriations to provide "seed monies" to communities in the form of grants and loans to aid them in financing certain types of cleanup actions. Although there was broad support for this effort, some questioned EPA's authority under CERCLA to use Superfund monies for the cleanup of these lower risk sites that were not listed on the NPL and that did not appear to warrant emergency removal actions under the Superfund program. Still, in the annual appropriations process, Congress set aside funding for brownfields cleanup assistance within the Superfund account for several years without specifically amending CERCLA for this purpose.

In the 107th Congress, Subtitle A and Subtitle C of Title II of the Small Business Liability Relief and Brownfields Revitalization Act of 2002 (P.L. 107-118, hereinafter referred to as the "Brownfields Act") amended CERCLA to provide explicit statutory authority for EPA to administer a Brownfields program separately from the Superfund program. The Brownfields Act authorized appropriations for this new program apart from appropriations for the Superfund account.[162] There had been some concern about the diversion of Superfund appropriations away from addressing the greater human health and environmental risks at NPL sites. Still, the portion of Superfund appropriations that had been spent on the cleanup of brownfields properties was relatively small compared to the total appropriation.

The program explicitly authorized in the Brownfields Act is similar in scope to the program that EPA had established in 1993, with the exception that the Brownfields Act allowed federal financial assistance for the cleanup of contamination resulting from releases of petroleum. As discussed earlier in the "Federal Response Authorities" section of this report, CERCLA otherwise does not apply to the cleanup of petroleum. The Brownfields Act also created two separate types of grants within the Brownfields program. One provides more direct financial assistance for the assessment and cleanup of individual properties. The other provides financial assistance to states and Indian tribes to aid them in carrying out their own cleanup programs, which in turn may assist in the cleanup of individual properties.

Specifically, Section 201 of the Brownfields Act amended Section 104 of CERCLA to add a new subsection (k) that authorized $200 million annually for grants to fund the assessment and cleanup of individual brownfields properties.[163] Entities generally eligible for these grants include state and local governments, Indian Tribes,[164] redevelopment agencies chartered or otherwise sanctioned by a state government, and land clearance authorities or other "quasi-governmental" entities operating under the supervision and control, or as an agent, of a local government. The grants are awarded on a competitive basis. The recipients may use the grant funds to characterize, assess, or remediate brownfields properties, or to capitalize revolving loan funds that in turn may finance the remediation of multiple brownfields properties by other entities, including loans issued to site owners or developers.

Section 231 of the act also added Section 128 to CERCLA, authorizing an additional $50 million annually for other grants to assist states and Indian Tribes in establishing or enhancing their own cleanup programs.[165] States and Tribes may use these monies to augment their own resources to assist with the cleanup of brownfields properties to prepare them for reuse. They also may use these monies to pursue the cleanup of other contaminated sites within their respective jurisdictions, which may present potential health or environmental risks but are not addressed under the federal Superfund program. The authorization of appropriations for both the Section 104(k) and Section 128 grants expired at the end of FY2006, but Congress has continued to fund these grants through the annual appropriations process without enacting reauthorizing legislation.

As discussed earlier in the "Limitations on Liability" section of this report, Section 222 of the Brownfields Act exempted "bona fide" prospective purchasers of contaminated properties from liability under CERCLA, if they satisfy the prerequisite statutory criteria. This exemption is intended to work in tandem with federal grants assistance under the Brownfields program to further the purpose of stimulating the economic redevelopment of contaminated properties. Accordingly, Section 104(k)(4)(B)(iii) of CERCLA[166] specifically authorizes the eligibility of bona fide prospective purchasers for brownfields grants. If a party cannot qualify for this exemption, or another exemption from liability, that party is not eligible to receive a brownfields grant.[167] This statutory prohibition on awarding Brownfields grants to potentially responsible parties is consistent with the policy premise of the liability scheme of CERCLA to hold the potentially responsible parties responsible for the costs of cleanup, so as to minimize the burden of these costs on the federal taxpayer who had no connection with the site.

Congress also has enacted certain tax incentives to encourage the cleanup of brownfields properties, through amendments to the Internal Revenue Code but not CERCLA itself. These incentives have constituted another form of federal financial assistance to support the cleanup of contaminated sites. Section 941(a) of the Taxpayer Relief Act of 1997 (P.L. 105-34) allowed a taxpayer to fully deduct the costs of cleaning up a brownfields property in the year the costs were incurred.[168] This type of deduction is referred to as "expensing," as opposed to "capitalizing" in which the costs would be deducted over a period of years. The tax incentive was intended to encourage property developers to rehabilitate sites where environmental contamination may be a deterrent to bringing nonproductive properties back into use. The tax deduction has no direct application for public sector entities, such as municipalities, which do not pay income taxes. Enacted in the 111th Congress, Section 745 of the Tax Relief, Unemployment Insurance Reauthorization, and Job Creation Act of 2010 (P.L. 111-312, Title VII, Subtitle C) extended this brownfields tax incentive through December 31, 2011.

Congress also had authorized another federal brownfields tax incentive in Section 702 of the American Jobs Creation Act of 2004 (P.L. 108-357, Title VII), which expired on December 31, 2009. This incentive addressed the treatment of gain or loss on the sale or exchange of certain qualified brownfields sites,[169] as defined in Section 101(39) of CERCLA.[170] This provision allowed a tax-exempt entity to invest in a qualified brownfields site, and not treat the gains as taxable "unrelated business income." To be eligible for this tax incentive, the entity must have incurred cleanup costs exceeding the greater of $550,000, or 12% of the property's fair market value in a remediated condition, in addition to meeting certain other requirements. The tax incentive was not available to parties who are potentially liable for the cleanup under Section 107 of CERCLA.

End Notes

[1] 42 U.S.C. § 9604(a).

[2] 42 U.S.C. § 9615.

[3] Executive Order 12580, Superfund Implementation, January 23, 1987, 52 *Federal Register* 2923. Hereinafter, references to Presidential authorities under CERCLA refer to those that have been delegated to EPA and other federal agencies, unless noted otherwise.

[4] 42 U.S.C. § 9603(a).
[5] 40 C.F.R. § 302.4.
[6] 42 U.S.C. § 9601(14).
[7] 42 U.S.C. § 9601(33).
[8] 33 U.S.C. § 2701 et. seq.
[9] 33 U.S.C. § 1321(c).
[10] 42 U.S.C. § 6991b(h).
[11] For EPA's interpretation of the statutory exclusion of releases of petroleum from the cleanup authorities of CERCLA, see Environmental Protection Agency, Office of General Counsel, *Scope of the CERCLA Petroleum Exclusion Under Sections 101(14) and 104(a)(2)*, July 31, 1987. Please note that Section 104(a)(2) of CERCLA, as originally enacted in 1980, defined the term "pollutant or contaminant." Section 101(f) of the Superfund Amendments and Reauthorization Act of 1986 re-designated the definition of this term in Section 101(33) of CERCLA, cited above.
[12] 42 U.S.C. § 9601(22).
[13] 40 C.F.R. §§ 302.4, Appendix B.
[14] 42 U.S.C. § 9607(i).
[15] 42 U.S.C. § 9607(j).
[16] 42 U.S.C. § 9604(a)(3).
[17] 40 C.F.R. Part 300.
[18] 42 U.S.C. § 9605(a).
[19] 42 U.S.C. § 9605(c).
[20] 40 C.F.R. 300.425(c)(3).
[21] For information on the number and status of sites across the United States listed on the NPL over time, see EPA's Superfund program website: http://www.epa.gov/superfund/sites/query/queryhtm/npltotal.htm.
[22] 42 U.S.C. § 9601(23).
[23] 42 U.S.C. § 9601(24).
[24] 40 C.F.R. § 300.425(b)(1).
[25] 42 U.S.C. § 9617.
[26] 42 U.S.C. § 9604(c)(1).
[27] 42 U.S.C. § 9604(c)(3).
[28] 42 U.S.C. § 9604(c)(6).
[29] 42 U.S.C. § 9621(a).
[30] 42 U.S.C. § 9621(b).
[31] 42 U.S.C. § 9621(c).
[32] 42 U.S.C. § 9621(d).
[33] 42 U.S.C. § 9621(d)(4).
[34] 42 U.S.C. § 300f et seq.
[35] 33 U.S.C. § 1313.
[36] 33 U.S.C. § 1314.
[37] 42 U.S.C. § 9621(f).
[38] 42 U.S.C. § 9621(f)(3).
[39] 42 U.S.C. § 9617.
[40] 42 U.S.C. § 9617(e).
[41] 42 U.S.C. § 9604(i).
[42] 42 U.S.C. § 9607.
[43] 42 U.S.C. § 9607(a).
[44] 42 U.S.C. § 9601(8).
[45] 42 U.S.C. § 9601(32).
[46] 33 U.S.C. § 1321.
[47] 42 U.S.C. § 9613(f)(1).
[48] 42 U.S.C. § 9607(f)(1).

[49] 42 U.S.C. § 9607(b).
[50] 42 U.S.C. § 9601(35).
[51] 42 U.S.C. § 9607(i).
[52] 42 U.S.C. § 9607(j).
[53] 42 U.S.C. § 9619.
[54] 42 U.S.C. § 9619(a)(1).
[55] 42 U.S.C. § 9619(a)(2).
[56] 42 U.S.C. § 9619(c).
[57] 42 U.S.C. § 9607(n).
[58] 42 U.S.C. § 9601(20).
[59] 42 U.S.C. § 9627.
[60] 42 U.S.C. § 9627(b).
[61] 42 U.S.C. § 9627(c).
[62] 42 U.S.C. § 9627(f).
[63] 42 U.S.C. § 9607(o).
[64] 42 U.S.C. § 9607(p).
[65] 42 U.S.C. § 9622.
[66] In 1989, EPA developed guidance for entering into Prospective Purchaser Agreements with potentially responsible parties under Section 122 of CERCLA. EPA revised its guidance in 1995 to expand the circumstances under which such agreements would be considered. See Environmental Protection Agency, Office of Enforcement and Compliance Assurance, *Guidance on Agreements with Prospective Purchasers of Contaminated Property*, May 24, 1995.
[67] 42 U.S.C. § 9607(r).
[68] 42 U.S.C. § 9601(40).
[69] 42 U.S.C. § 9607(q).
[70] 42 U.S.C. § 9601(35).
[71] 42 U.S.C. § 9607(b)(3).
[72] 42 U.S.C. § 9601(35)(B).
[73] 40 C.F.R. Part 312.
[74] 42 U.S.C. § 9601(35)(B)(i)(II).
[75] 26 U.S.C. § 4611.
[76] 26 U.S.C. § 4661.
[77] 26 U.S.C. § 4671.
[78] 26 U.S.C. § 59A.
[79] 26 U.S.C. § 9508. For a discussion of the Leaking Underground Storage Tank Trust Fund, see CRS Report RS21201, *Leaking Underground Storage Tanks (USTs): Prevention and Cleanup*, by Mary Tiemann.
[80] 26 U.S.C. § 9509. For a discussion of the Oil Spill Liability Trust Fund, see CRS Report RL33705, *Oil Spills in U.S. Coastal Waters: Background and Governance*, by Jonathan L. Ramseur.
[81] 42 U.S.C. § 9622(b)(3).
[82] For background information on the use and overall status of Superfund Special Accounts, see EPA's Office of Enforcement and Compliance Assurance website: http://www.epa.gov/oecaerth/cleanup/superfund/spec-acct.html.
[83] 42 U.S.C. § 9606(a).
[84] 42 U.S.C. § 9606(b)(1).
[85] 42 U.S.C. § 9607(c)(3).
[86] 42 U.S.C. § 9606(b)(2).
[87] 42 U.S.C. § 9607(a).
[88] 42 U.S.C. § 9604(i).
[89] 42 U.S.C. § 9613(g)(2).
[90] 42 U.S.C. § 9622.

[91] 42 U.S.C. § 9622(f).
[92] 42 U.S.C. § 9622(b)(3).
[93] 42 U.S.C. § 9613(f)(2).
[94] 42 U.S.C. § 9609.
[95] Environmental Protection Agency, Office of Enforcement and Compliance Assurance, *General Policy on Superfund Ability to Pay Determinations*, September 30, 1997. The full text of this document is available on the agency's website: http://www.epa.gov/compliance/resources/policies/cleanup/superfund/genpol-atp-rpt.pdf.
[96] For example, see United States v. Bay Area Battery, 895 F. Supp. 1524 (N.D.Fla. 1995).
[97] 42 U.S.C. § 9622 (g)(7)(A).
[98] 42 U.S.C. § 9622(g)(7)(B).
[99] 42 U.S.C. § 9622(g)(7)(C).
[100] 42 U.S.C. § 9622(g)(8)(A).
[101] 42 U.S.C. § 9622(g)(8)(C).
[102] 42 U.S.C. § 9622(g)(12).
[103] 42 U.S.C. § 9622(g)(8)(B).
[104] 42 U.S.C. § 9622(g)(9).
[105] 42 U.S.C. § 9622(g)(11).
[106] Environmental Protection Agency, Environmental Appeals Board, *The Environmental Appeals Board Practice Manual*, June 2004, "EAB Jurisdiction," pp. 2-4.
[107] 42 U.S.C. § 9613(f)(1).
[108] 42 U.S.C. § 9613(f)(3)(B).
[109] 42 U.S.C. § 9613(f)(2).
[110] 42 U.S.C. § 9659.
[111] 42 U.S.C. § 9613(h)(4).
[112] 42 U.S.C. § 9659(a)(1).
[113] For the purpose of citizen suits, Section 310(a)(1) states that "any person" includes "the United States and any other governmental instrumentality or agency, to the extent permitted by the 11th amendment to the U.S. Constitution."
[114] 42 U.S.C. § 9659(b)(1).
[115] 42 U.S.C. § 9659(c).
[116] 42 U.S.C. § 9659(d)(1).
[117] 40 C.F.R. Part 374.
[118] 42 U.S.C. § 9659(d)(2).
[119] 42 U.S.C. § 9659(a)(2).
[120] 42 U.S.C. § 9659(b)(2).
[121] 42 U.S.C. § 9659(c).
[122] 42 U.S.C. § 9659(e).
[123] 40 C.F.R. Part 374.
[124] 42 U.S.C. § 9659(a), 9660.
[125] 42 U.S.C. § 9601(21).
[126] 42 U.S.C. § 9620.
[127] 42 U.S.C. § 9611(e).
[128] 42 U.S.C. § 9611(e)(3).
[129] The vast majority of the funds are appropriated to the Department of Defense and the Department of Energy for the cleanup of federal facilities which served national defense purposes.
[130] See General Accounting Office, *The Judgment Fund and Litigative Awards Under the Comprehensive Environmental Response, Compensation, and Liability Act*, B-253179, November 29, 1993, available on GAO's website: http://archive.gao.gov/lglpdf63/151167.pdf.
[131] 31 U.S.C. § 1304.
[132] 42 U.S.C. § 9620(e).

[133] 42 U.S.C. § 9620(f).
[134] 42 U.S.C. § 9620(e)(1).
[135] 42 U.S.C. § 9620(e)(2).
[136] 42 U.S.C. § 9620(e)(4).
[137] 42 U.S.C. § 9620(e)(4)(A).
[138] 42 U.S.C. § 9620(g).
[139] 42 U.S.C. § 9620(e)(5).
[140] 42 U.S.C. § 9620(e)(3).
[141] 42 U.S.C. § 9620(f).
[142] 42 U.S.C. § 9621(f).
[143] 42 U.S.C. § 9621(d).
[144] 42 U.S.C. § 9621(f)(3).
[145] 42 U.S.C. § 9620(a)(4).
[146] 42 U.S.C. § 6901 et. seq. The Solid Waste Disposal Act often is referred to as the Resource Conservation and Recovery Act (RCRA; P.L. 94-580), which substantially amended the Solid Waste Disposal Act in 1976 to regulate the storage, treatment, and disposal of hazardous waste.
[147] The Hazardous and Solid Waste Amendments of 1984 (P.L. 98-616) amended the Solid Waste Disposal Act to require operators of hazardous waste facilities to perform corrective actions to clean up environmental contamination resulting from the improper management or disposal of hazardous wastes.
[148] 42 U.S.C. § 9620(i).
[149] 42 U.S.C. § 9620(h).
[150] 42 U.S.C. § 9620(h)(3).
[151] 42 U.S.C. § 9620(h)(4).
[152] 42 U.S.C. § 9620(h)(4)(D).
[153] 42 U.S.C. § 9620(h)(3)(C).
[154] 10 U.S.C. § 2701 et. seq.
[155] Because of its specific applicability to U.S. military facilities, Section 211 of the Superfund Amendments and Reauthorization Act of 1986 is codified in Title 10—"Armed Forces"—of the *United States Code*, rather than Title 42—"Public Health and Welfare"—under which the provisions of CERCLA generally are codified.
[156] 10 U.S.C. § 2701(c).
[157] 10 U.S.C. § 2710.
[158] 10 U.S.C. § 2703(b).
[159] 10 U.S.C. § 2710(d). The statutory scope of the Military Munitions Response Program also specifically excludes any locations outside the United States, the presence of military munitions resulting from combat operations, and operational munitions storage and manufacturing facilities. However, the operation of storage and manufacturing facilities in the United States may be subject to regulation under other statutory authorities.
[160] 40 C.F.R. Part 266, Subpart M.
[161] 42 U.S.C. § 9620(j).
[162] Since the enactment of the Brownfields Act, Congress has appropriated specific levels of funding for Brownfields grants within EPA's State and Tribal Assistance Grants account, and has appropriated funds to administer these grants within EPA's Environmental Programs and Management account.
[163] 42 U.S.C. § 9604(k).
[164] In Alaska, Tribal eligibility is extended specifically to the Alaska National Regional Corporation and Alaska Native Village Corporation as defined in the Alaska Native Claims Settlement Act. The Metakatla Indian Community also is specifically authorized in the statute as being eligible for the grants.
[165] 42 U.S.C. § 9628.
[166] 42 U.S.C. § 9604(k)(4)(B)(iii).

[167] 42 U.S.C. § 9604(k)(4)(B)(i)(IV).
[168] 26 U.S.C. § 198.
[169] 26 U.S.C. § 512(b)(19).
[170] 42 U.S.C. § 9601(39).

INDEX

D

E

F

G

H

I

J

L

M

N

O

P

Q

R